WHAT ON EARTH?

Student Manual

DEVELOPED BY
EDUCATION DEVELOPMENT CENTER, INC.

KENDALL/HUNT PUBLISHING COMPANY
4050 Westmark Drive P.O. Box 1840 Dubuque, Iowa 52004-1840

This book was prepared with the support of National Science Foundation (NSF) Grant ESI-9255722. However, any opinions, findings, conclusions and/or recommendations herein are those of the author and do not necessarily reflect the view of NSF.

Library of Congress Catalog Card Number: 96-80279

ISBN 0-7872-2210-0

Printed in the United States of America
10 9 8 7 6 5 4 3 2 1

EDC Education Development Center, Inc.

CENTER FOR SCIENCE EDUCATION

Dear Students:

Welcome to *Insights in Biology*. This module, *What on Earth?* explores the interrelationships and interactions among organisms and their environments. Throughout the module, the questions "What relationships exist among organisms? How do organisms interact with their environments? How did the vast diversity of organisms we see on Earth come to be? What might the future hold both for the organisms on Earth and their habitats?" guide the introduction and development of the concepts of ecology. You will find that the principles that influence the continuance of life on Earth are more complex than you imagined. Ecology serves as the basis for examining the environmental problems that face us. Their possible solutions must not be left only to scientist or to industrialists; educated citizens must be involved in deciding how to maintain the earth and its resources, including living things.

Glance through the pages of this manual. Your first instinct is correct—this is not a traditional biology textbook. Although textbooks provide a good deal of useful information, they are not the only way to discover science. In this Student Manual, you will find that chapters have been replaced by Learning Experiences with activities and readings. The activities include laboratory experiments, concept-mapping, model-building, simulation exercises, and case studies. These learning experiences emphasize the processes of science, the connections among concepts in ecology, and the development of critical-thinking skills.

One of our main goals is to engage you in the excitement of biology. It is a discipline that is as alive as the subject it portrays; new questions arise, new theories based on evidence are proposed, and new understandings are achieved. As a result of these new insights, decisions are made that will impact your everyday lives. We hope that this curriculum encourages you to ask questions, to develop greater problem-solving and thinking skills, and to recognize the importance of science in your life.

Insights in Biology Staff

55 CHAPEL STREET
NEWTON, MASSACHUSETTS 02158-1060
TELEPHONE: 617-969-7100
FAX: 617-630-8439

Table of Contents

Learning Experiences

Appendix

Home Is Where the Habitat Is

PROLOGUE The natural world surrounds us, but we are often oblivious to it. Think about your journey to school today. What living things did you see? Why are those particular organisms living there? How and where do they get the resources to stay alive? In what ways do these organisms interact with one another and with the environment? Living things are linked to their habitat by a multitude of factors. These factors include the nonliving or physical conditions as well as the number and kinds of organisms present in the area. The science of *ecology* focuses on the interactions among living things and their relationships to the environment.

As you explore the concepts in this module, your responses to these questions—and other questions of your own—may change. The activities, investigations, and readings in this module will deepen and expand your understanding of the natural world.

In this learning experience, you will begin an exploration of ecological systems (ecosystems) and the organisms that inhabit them. You will observe an area in your school yard, identify both the organisms that live there and other less visible components that are present in the *habitat*. Using these observations and your analysis of interactions among organisms, you will then design and create a mini-ecosystem which will be used as a model for the principles you will study in this module. Finally, you will be asked to consider the political, economic, and environmental issues surrounding the reintroduction of wolves into areas where they once lived and speculate how replacing an organism might affect the structure or function of that ecosystem.

LIFE IN A SQUARE METER

INTRODUCTION All around you are living things: grass, weeds, trees, birds, insects and other forms of life. In any environment, each plant or animal lives in its own special place, its habitat, and carries out its life functions. It uses its environment to obtain what it needs and, in turn, has effects on the environment. In this investigation, you will observe the organisms that live above and below the ground within a square-meter area. What are these organisms? How do the plants and animals obtain food? Where are the inorganic materials they need, and how do they obtain them? As you think about these questions you will begin to consider the relationships of organisms to the environment within an ecosystem.

MATERIALS NEEDED

For each group of four students:

- 1 large bag or box for carrying materials
- 1 meter stick or tape measure
- 4 large nails or golf tees
- 1 hammer or mallet or fist-sized stone
- 5 m of string
- 1 thermometer
- 1 trowel, spade, 5-cm pipe or small bulb planter
- 1 sheet of graph paper
- 1 sheet of newspaper
- 2 hand lenses or magnifying glasses
- 1 petri dish
- 1 soil test kit (optional)

PROCEDURE

1. Designate a recorder and an illustrator for your group.
2. Collect the materials for your group and follow your teacher outdoors.
3. Choose an area that is as diverse as possible; that is, an area that might have a tree or shrub, a variety of weeds, and grass. This area should not be immediately next to another group's area, but must be within the sound of your teacher's voice.
4. Hammer one nail (or tee) into the ground. Measure one meter from this nail and hammer in a second nail. Continue in this way until you have a square with a nail at each corner.

***CAUTION:** Do not step in your square-meter plot or that of another group.*

5. Attach string to the four nails to rope off your square-meter plot (see Figure 1.1).
6. Have the illustrator draw a top view of the plot to scale on graph paper. Label the plant life, soil covering and debris such as rocks, leaves, or pine needles, and any other features of your square meter. Write all names of students in the group at the bottom of the drawing.
7. In your notebook, record the date, time, temperature, weather conditions and the location of your plot.
8. Create a data chart to record the following: organisms (plant and animal) observed, their locations, and their numbers (or percentage of the plot they occupy).
9. Kneel next to one side of your area and carefully look at each type of plant. Record each plant name in your data chart. If you cannot identify a plant, draw a picture of its shape and leaves.
10. Continue and carefully look for each type of animal in your area. Record in your data chart. If you cannot identify an organism, draw it (if applicable, include the number of legs it has).
11. Spread a sheet of newspaper on the ground outside the square-meter area.
12. Dig straight down into the soil in one part of your plot using a trowel (or use a 5-cm pipe or small bulb planter) to get a soil core. Remove soil intact and place on newspaper. Use a hand lens or magnifying glass to note and count all the organisms (including any eggs) living in the soil. Record in your data chart. Note any soil layers you see.
13. Place the lower part of the thermometer in the hole, burying the bulb if possible. Let it stand in the hole for 3–5 minutes to adjust to the soil temperature. Record the temperature.
14. If a soil test kit is available, test the pH of the soil according to the directions in the kit. Record in your data chart.
15. Replace the soil.
16. Return all materials to the classroom.

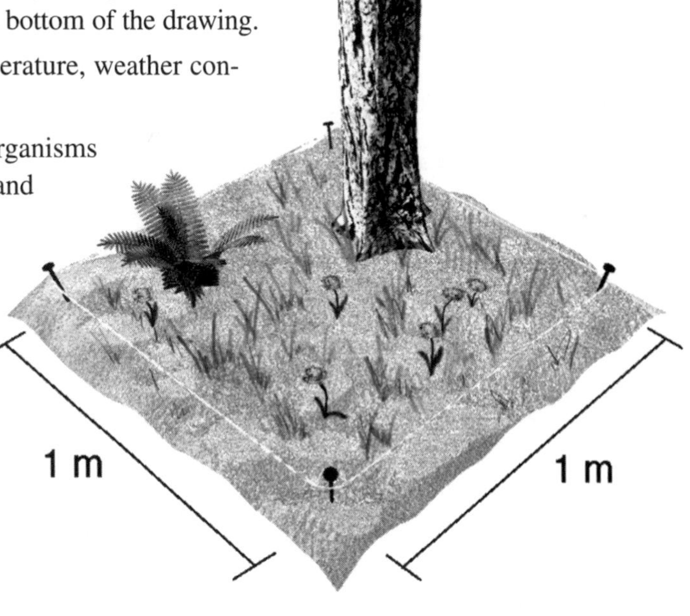

Figure 1.1
Square-meter plot.

ANALYSIS

Write responses to the following in your notebook.

1. How do the plants and animals in your plot relate to one another?
2. How do nonliving factors affect the plot and the organisms in it?

3. Is your square-meter plot considered an ecosystem or part of an ecosystem? Explain your response.
4. In a short paragraph, describe what information you might need to understand the connections among the organisms and their environment better?

READING

THE FACTORS OF LIFE

THE LIFE OF A POND

Beneath the surface of the water the amount of light lessens, causing the colors to darken rapidly from a light blue at the top to a dark green near the bottom. Thin light-colored stems bearing long, slender leaves reach toward the surface. A school of small, black-banded silvery fish swim through these stems. A large insect uses its long pointed legs to grasp one of the fish, which struggles and then is still. Tiny shrimp-like animals swim through the water, and a large number of much smaller living things dart and tumble around them. On the mud at the bottom are worms and other creatures with sprawling legs.

Such a naturally occurring group of organisms living in a particular area, depending on and sustaining each other, is termed a *biotic* (living) community. A community is influenced by and dependent upon *abiotic* (nonliving) factors such as sunlight, soil, topography, wind, temperature, moisture, and minerals. The combination of the biotic and abiotic factors creates an *ecosystem.* The pond described above is an example of an ecosystem. So intricately knit is the web of interacting factors within an ecosystem that, should one factor change, all the rest of the relationships may be affected to some degree.

Figure 1.2
Cycle of life in an ecosystem.

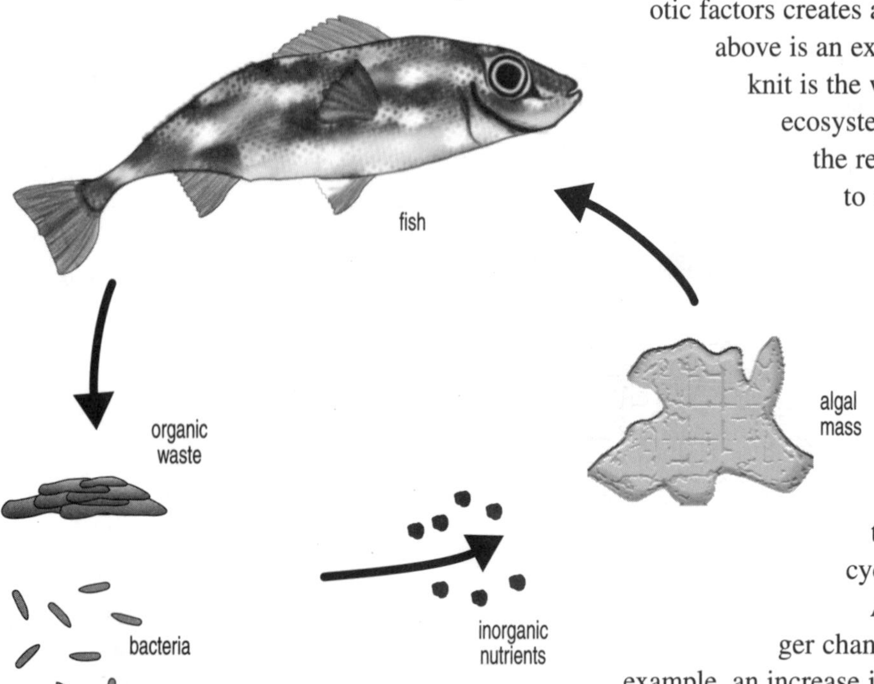

In a pond, for example, some fish eat algae and excrete organic waste; bacteria then use this waste as a nutrient and break it down to inorganic materials. Algae use the inorganic materials for making food. Thus, each population within the community is affected by the activities of the other organisms. It is a cycle of life and death (see Figure 1.2).

A change in an abiotic factor can trigger changes throughout the ecosystem. For example, an increase in available nitrogen can lead to overgrowth of algae. Too much algae in the water can block the passage of sunlight, and without sunlight the algae cannot make food. They will die, sink to the bottom, and become organic waste fed on by bacteria. The resulting explosive growth of the bacterial population will deplete the amount of oxygen in the water. While the fish may temporarily increase in number in response to excess food—the algae—the low oxygen may eventually lead to the death of the fish.

THE MATTER OF LIFE

At the core of every organism's interaction with its environment is the organism's need for energy and nutrients—energy to power its life processes, and raw materials to build and maintain living tissue. In a continuous cycle, plants and animals exchange the chemicals necessary for building materials.

Organisms in an ecosystem can be named according to how they obtain food. Those organisms capable of making their own food (*autotrophs*) manufacture sugar from solar energy and simple chemicals found in the environment through the process of photosynthesis. Organisms that get their energy by consuming other organic matter are called *heterotrophs*. In order for life to cycle, a third group of organisms, the *decomposers*, are necessary to break down organic material into inorganic materials that autotrophs will use to build biomolecules.

In an ecosystem, the biotic and abiotic factors are interconnected; and a shift in any of the factors will affect some part of the community. In healthy ecosystems, these changes are self-correcting. Only sudden and/or extreme change can disrupt the interactions in an ecosystem and lead to serious consequences.

ANALYSIS

Write responses to the following in your notebook.

1. Explain three different ways in which organisms in an ecosystem are dependent upon the abiotic factors in the environment.
2. List the organisms in the opening paragraph of this reading. Decide whether each is an autotroph, a heterotroph, or a decomposer. Explain your decision.
3. Explain how autotrophs, heterotrophs, and decomposers interact to maintain an ecosystem.
4. Based on your understanding of ecosystems, is a tree considered an ecosystem? Why or why not?

Life in a Jar

INTRODUCTION What can a jar full of mud, leaves, and water tell us about life on Earth? Can you create an ecosystem in a jar that mimics how living things interact with their environment? In this activity, you will develop a small ecosystem (called an "ecocolumn") that will serve as a model for the interactions within any ecosystem. What would the jar need to contain? What are the important components necessary for this miniecosystem?

MATERIALS NEEDED

For each group of four students:

- 2 or more clear plastic 2-L soft drink bottles of the same brand, with caps
- 1 wax marking pencil
- 1 razor blade or scalpel
- 1 scissors
- 1 awl or nail
- clear waterproof tape
- 1 darning needle or large safety pin
- silicone sealant
- soil, water, plants, compost, fruit flies, spiders, snails and/or other small organisms
- wick (cotton rope or other absorbent material)

For the class:

- 2 or 3 large shoebox tops

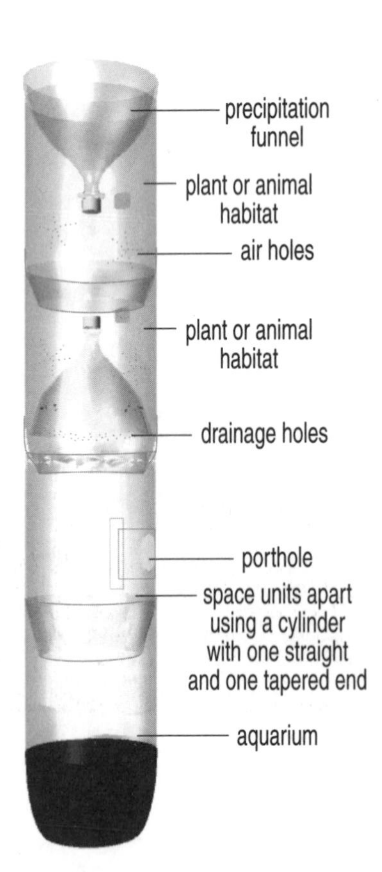

Figure 1.3
An example of a multichambered ecocolumn. Create one or more habitats for your own column. *(Ecocolumns were developed by the Bottle Biology Project, Department of Plant Pathology, College of Agricultural and Life Sciences, University of Wisconsin—Madison.)*

PROCEDURE

1. **STOP & THINK** What type of ecosystem do you wish to model? Determine the number of chambers (providing one or more habitats) you would need to create a functioning ecocolumn. See Figure 1.3 for one idea about creating and filling chambers.
2. **STOP & THINK** What abiotic components will the ecosystem need for maintaining life? What materials will be needed in the ecocolumn to represent these components?
3. **STOP & THINK** Determine the materials and specimens you will need to collect from your neighborhood in order to create the ecocolumn.

4. Draw a plan of the chambers and list the organisms and other materials you plan to include in each habitat.
5. Remove labels from each of the bottles as follows: Fill the bottle 1/4 full with hot water (50°–65°C), making sure it is not too hot (see Figure 1.4a). Screw the cap back on the bottle in order to retain pressure inside. Tip the bottle on its side so that the water warms the area where the label is attached to the bottle. Gently peel off the label (see Figure 1.4b).

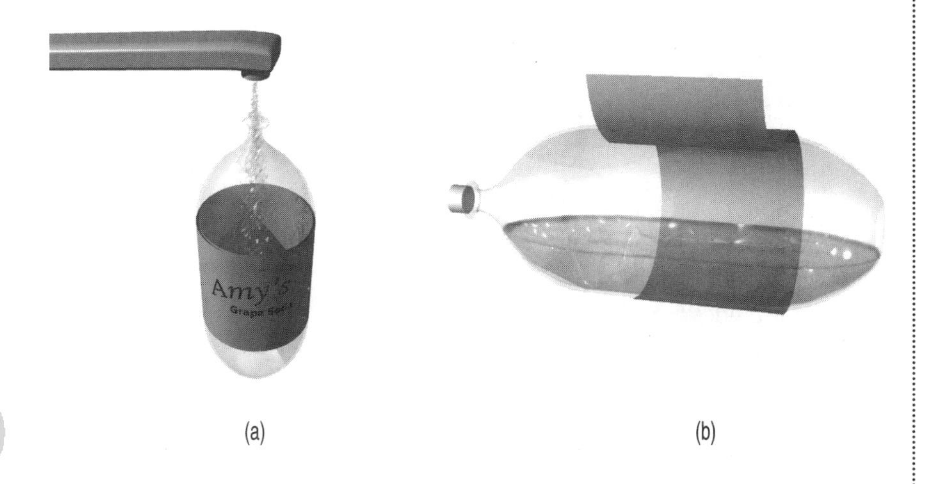

Figure 1.4
a) Add hot water. b) Tip bottle on its side, so the water warms and softens the glue.

6. Mark and cut the bottles to create the chambers as described in Figure 1.5 below.

CAUTION: *Be careful when handling razor blades and scissors; they are extremely sharp. Always cut away from your body, and hold the bottle on a stable surface.*

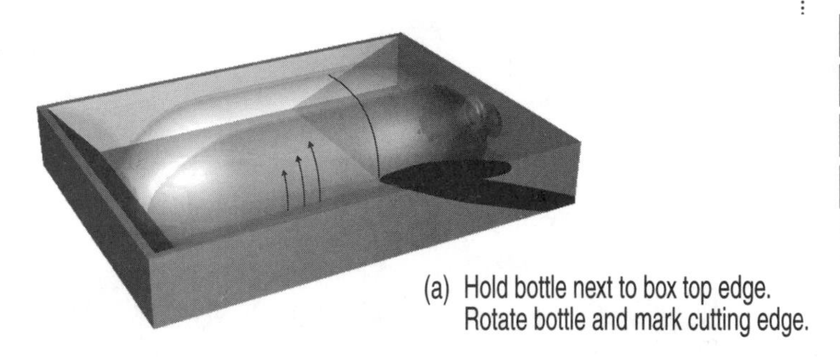

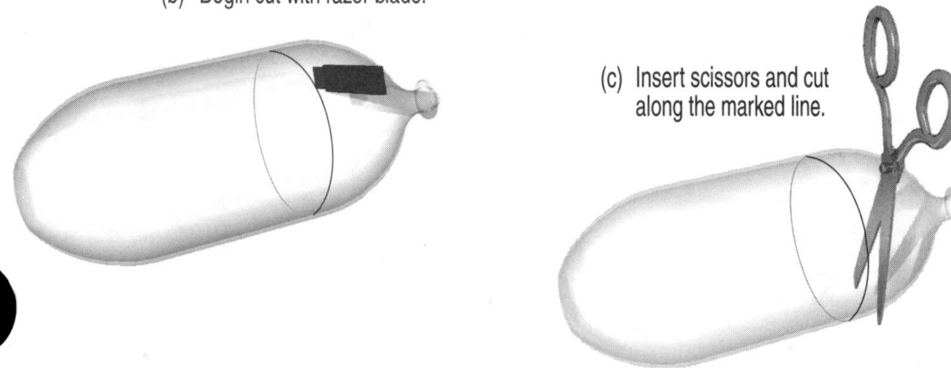

Figure 1.5
Cutting bottles for ecocolumn chambers.

CAUTION: These bottles can become top-heavy and may tip. Provide support for them.

7. Connect the chambers by punching holes in the caps and bottle tops with an awl. Tape together the chambers that are to be left undisturbed. Poke air holes in any chambers that contain living organisms. Make sure any chamber with water is leakproof.
8. Determine the number and height of the drainage holes. These will affect the environment in a soil-filled chamber.
9. Add the biotic and abiotic components to the appropriate chamber. Record these and their placement in your notebook.
10. Observe your ecocolumn daily. Note any changes and describe them
11. STOP & THINK Determine if and when you will need to add the food or water and keep a record in your notebook. Are there other variables you need or would like to keep track of during the investigation?
12. At the conclusion of the investigation, write a report that includes the following:
    - a labeled diagram of the ecocolumn including the organisms as well as the abiotic materials you placed in each chamber
    - a description of the changes you observed
    - a data chart that summarizes the data you collected, any trends you observed, and notation of when organisms were added and when organisms appeared to die
    - a diagram of food chains or food webs within the ecocolumn
    - an analysis of your observations and data using concepts from the module
    - conclusions you were able to draw from the investigation

ANALYSIS

Write responses to the following in your notebook.

1. How is your ecosystem similar to a large, natural ecosystem on Earth?
2. If you placed a cap on the ecocolumn and omitted the air holes—thus creating a closed system—how would it be similar to and different from Earth?
3. Is it possible for life to be maintained indefinitely in an aquarium containing plants, animals, and decomposers, and having a sealed glass top? Explain your response.

Your report will be assessed on the following criteria:

- Does it include a labeled diagram?
- Have you gathered sufficient data and observations? Are they included in the report.
- Are food chains or food webs included?
- Is there an explanation of why the ecocolumn could or could not continue to flourish?

CASE STUDY

Wolves

AFTER 50 YEARS, FABLED PREDATORS ARE BACK HOME

Excerpted from Michael Milstein, The Boston Globe. *Monday, January 23, 1995, pp. 25–26.*

HINTON, Alberta, Canada—...Fifty years after the gray wolf was poisoned out of existence—with the aid and encouragement of the US government—federal biologists were returning the storied predator to its original place in two wild remnants of the American West.

In Idaho, the return came on January 14, when four beleaguered animals left their crates timidly and then darted into the thick timber of one of the largest roadless regions in the West.

In Yellowstone two days earlier, the wolves had emerged over several hours, cautiously stepping out into one-acre, chain-link pens, where they will stay for six to eight weeks. Biologists hope that transition period will persuade them to become permanent American immigrants, rather than high-tail it back to Canada.

Stretching their stiff legs after the long trip, the wolves nibbled on elk and deer meat left for them, paced along the fences of their pen and tugged at the heavy wire, seeking a way out.

In Canada, meanwhile, wildlife workers were pursuing the rest of the 30 animals that will be brought south and turned loose this winter—15 each in Yellowstone and Idaho. The intent is to capture and release 30 animals annually over the next three to five winters. By then, biologists say, Yellowstone and

Idaho should each have 10 fruitful wolf packs, or about 100 animals, enough to sustain their population but few enough to be controlled by regulated hunting.

The wolves' return to Yellowstone fills a missing patch in the diverse biological quilt of the world's first national park. Not only do wolves belong there, park biologists say, but the sly hunters provide a check on the thousands of elk, deer, and bison that in places eat the spring grasses down to bare ground.

"This extraordinary creature creates for us a complete portrait of what a national park should be," said Interior Secretary Bruce Babbitt, noting that his own ranching ancestors had joined in the systematic poisoning and trapping that erased wolves from Yellowstone and the West early in the century.

Since then, lone wolves have occasionally wandered into Yellowstone, but none stayed. In the lower 48 states, wolves lived only in Minnesota and far northern Montana.

The wolves' return to Yellowstone came slowly. As far back as 1980, the US Fish and Wildlife Service, overseer of endangered species, proposed transplanting wolves to the 2.2-million-acre park. But the plan was repeatedly stalled by political wrangling, by studies costing $6 million so far and by a series of legal challenges that continued even after the first of the wolves had been captured.

Along the way, the Fish and Wildlife Service received 160,000 public comments, the most ever generated by any federal action. Most wanted the wolf back. And so early this month, sharp-shooting biologists armed with tranquilizer guns and riding in helicopters scoured Canadian forests so cold that fumes from pulp plants hung in valleys like soup in a bowl. One wolf died when a dart accidentally punctured its lung, but otherwise the capture went smoothly. . .

Soon after Yellowstone Park was established in 1872, hunters—with government encouragement—began lacing

Continued on next page

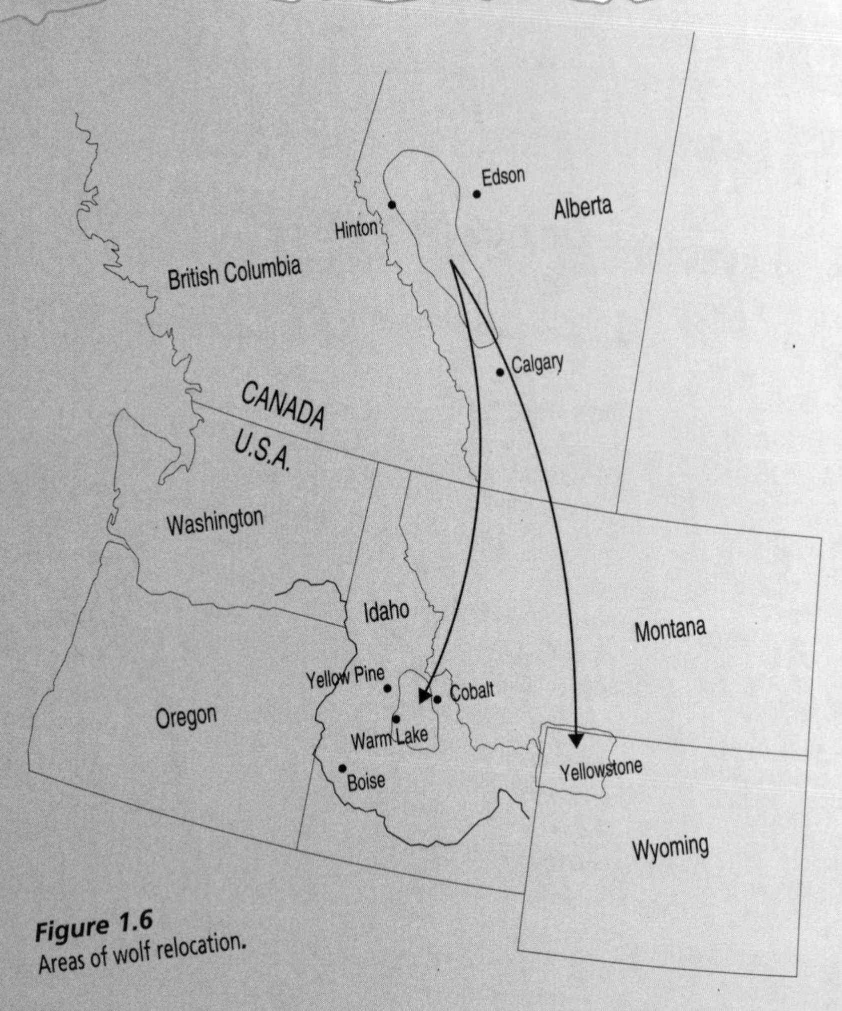

Figure 1.6
Areas of wolf relocation.

elk carcasses with strychnine to poison wolves that preyed on the majestic elk and bighorn sheep that tourists were coming to see. Such was the loathing for wolves that early eradication efforts included shoving explosives into dens of newborn pups. From 1914 to 1926, park rangers killed at least 136 wolves.

"Gray wolves are increasing and have become a decided menace to the herds of elk, deer, mountain sheep and antelope," a park superintendent reported in 1915. "An effort will be made the coming winter to capture or kill them."

The effort took several decades; the last known Yellowstone wolf was killed in 1944. And while the government has changed its thinking in the years since, many ranchers and farmers have not. The coyotes, cougars, bears and eagles that stalk their livestock are predators enough, they say, but they concede that they are in the minority.

"We lost the battle," said Stan Flitner, who grazes cattle near Shell [Wyoming], 150 miles from Yellowstone's boundary. "It seems to be the law of the land that we've got to reintroduce an animal that's going to cause nothing but problems."

Biologists agree that a healthy wolf population might kill perhaps 20 cattle and 70 sheep in a year. But studies suggest that wolves could attract millions of new tourist dollars, more than offsetting the economic loss to ranchers or hunters that compete with wolves for game. . .

To ease any hardships,

Defenders of Wildlife has created a private fund to compensate ranchers who lose livestock to wolves. So far, more than $100,000 has been collected, part of it in contributions from children. The fund has paid out $17,000 for cattle and sheep killed by the few wolves that have migrated from Canada into northwest Montana.

"If there's a price to making sure future generations get to see a wolf, making its living where it belongs, we're willing to foot it," said Defenders president Rodger Schlickeisen.

The opponents, however, have not given up. Even while wolves prowled their pens in Yellowstone, the Wyoming legislature considered a symbolic bill that would put a $500 bounty on the federally protected animals. In another last-ditch effort, the American Farm Bureau Federation last month sued the government to stop the reintroduction, and now, to send any released wolves back to Canada. . .

"It's unfortunate our people have to bear this burden while everyone else gets this warm fuzzy feeling," said Jake Cummins, a rancher and vice president of the Farm Bureau's Montana chapter.

Instead of attacking livestock, though, biologists expect that the wolves will prey mainly on elk, bison, and other big herbivores that have overrun Yellowstone in the absence of the predators.

"If these wolves realize they've got a buffet sitting right here, they'll have no reason to go anywhere else," said Yellowstone biologist Michael Phillips, who will watch over the park's newest residents.... Computer models foresee wolves culling Yellowstone's 40,000 elk by 5-to-30 percent. That influence may well be healthful, limiting elk numbers that now swing wildly from boom to bust and leave thousands to die during hard winters.

Wolves should also stabilize the growing herds of bison, which now are shot if they leave the park.

In the absence of wolves, coyotes have taken over [parts of their] ecological niche. Wolves should retake their position atop the food chain, driving off the coyotes and leaving more room for the foxes and raptors that compete with coyotes.

"It may be hard on coyotes," said Robert Crabtree, an independent ecologist specializing in coyote behavior, "but wolves may be the critical link that increases diversity and gets this ecosystem back on track."

ANALYSIS

Write responses to the following in your notebook.

1. What results followed the removal of wolves from the Yellowstone ecosystem?
2. The reintroduction of the wolves to Yellowstone is controversial. Identify at least two important issues on each side of the controversy.
3. If you had been in charge of this project, what decision would you have made about whether to introduce the wolves into Wyoming and Idaho? Discuss the reasons for your decision.
4. List four important values you hold that influenced your decision and explain how they influenced you.

EXTENDING IDEAS

- Read *The Return of the Wolf to Yellowstone* by Thomas McNamee (NY: Holt, 1997) and write an essay describing how the presentation of the issues in the book affected your own point of view and informed you about ecology and human nature.
- Research the ecological and political issues that surround other species, such as the peregrine falcon, the California condor, or the bison. What factors determined their reintroduction to their once-native habitats and what have been the results?

ON THE JOB

PARK RANGER Search planes flew low over the trails near the top of the mountain. Groups of trained individuals trudged up the tough incline. All were in search of two teenagers who had separated from their school hiking group and had not returned to the base camp. There was still daylight by which to search, but snow was expected overnight and winds near the summit were known to reach 70 mph and beyond—conditions that could surely be fatal for two youngsters not prepared for an overnight stay.

Curtis was a park ranger with the National Park Service. Although lost hikers were not an everyday occurrence on the popular mountain, it happened often enough for him to have to be trained in survival techniques and be physically fit. He enjoyed being outdoors and knew his job was important for keeping natural, recreational areas such as this safe for travelers.

Curtis grew up in the middle of a big city. His apartment complex was surrounded by concrete and steel and his experiences with wildlife were limited to occasional squirrel sitings and numerous encounters with stray dogs and cats. Curtis longed for trees and ponds, frogs and snakes. He knew that when he grew up, he would work outdoors to be as close to nature as possible.

A boy scout growing up, Curtis had his sites set on learning everything he could about the outdoors. After high school, he chose to go to a community college although he knew that a person need only have a high school diploma, a driver's license, and experience or knowledge of the outdoors to start their career as a park ranger. In college, he concentrated in recreation and tourism, while taking general and field science courses. The summer after his first year, he worked as a camp counselor. He also became a member of the Sierra Club, taking trips to different sites for hiking, canoeing, and skiing. When he graduated, he began a six month training program with the National Park Service, eventually leading to his job at the popular hiking and camping site.

No day was typical for Curtis, though some were less strenuous than others. There were days when he stayed at the base camp, talking with hikers before and after their climbs. He was available for questions and was sure to warn visitors of any particular dangers they might encounter while hiking. When not in base camp, Curtis could usually be found hiking the trails. He would direct people to trails that were, perhaps, easier or more challenging. He was watchful that everyone was respecting the natural areas, not doing anything that might endanger others such as starting fires or, in some cases during the snowy season, causing avalanches. If he encountered someone acting suspiciously or causing destruction to the trees or other areas of the mountain, he was responsible for taking charge of the situation in the safest way possible. His job was to be aware of the land and people in his surroundings.

Curtis was always on call for emergencies, and this day was one of them. After searching the mountain for several hours, the students were found near the base of a ravine, cold and hungry, but all right. Another day with a happy ending.

Oh, What a Tangled Web

PROLOGUE Did you know that there may be up to two million earthworms under a football field? Or that vacant lots are not just areas filled with dirt, litter, and broken glass, but can be the home of countless plants and animals? Even your school yard, no matter how barren it may appear, supports life. What living things are out there? What are their roles, and how do they interact with each other? Why are they there and not in some other type of environment?

In this learning experience, you will explore what organisms live in a specific environment and what roles they play there. You will then examine the complex feeding relationships among organisms in a marsh ecosystem.

Mystery Soil

INTRODUCTION Organisms do not live in isolation, but exist with biotic and abiotic factors from which they derive the materials needed for life. All the interacting populations of plants, animals, and microorganisms in an environment make up a *community* of organisms. The interactions between the community and the abiotic factors in that area create an ecosystem.

Each population of organisms performs a role within their community. All the things they do—where they live, what they eat, what eats them, and how they interact with biotic and abiotic factors—is referred to as the organism's ecological *niche.*

In this activity, you will observe a soil sample, collect data, and determine the niche of each type of organism you see. As you proceed, you will need to think like an ecologist in order to begin to understand this community of living organisms and to ask: Is the soil an ecosystem? What lives there? What do these organisms do? What do they need to live? What interactions occur in soil?

MATERIALS NEEDED

For each group of four students:

- 1 zippered plastic bag containing a sample of soil
- 2 hand lenses or magnifying glasses
- 2 tweezers
- 2 dissecting needles (or coffee stirrers)
- 1 sheet of newspaper
- 1 petri dish
- dissecting microscope
- rotting log (if available)

For the class:

- invertebrate field guides (optional)

PROCEDURE

CAUTION: Waft the scent from the bag to your nose by waving your hand from the soil to your nose. Do not smell the soil directly. When you finish working with the soil, wash your hands.

1. Spread a sheet of newspaper on your laboratory table. Place the bag containing the soil sample on it. Do not shake or mix the soil.
2. Open the bag and smell the soil. Write a description of the soil smell in your notebook.
3. Observe the soil's color, texture, and any organisms it contains. Record your observations in your notebook.
4. Use the tweezers and needles to probe through the soil, being careful not to mix the layers. Take notes on all organisms seen. (If you can not name an organism determine whether it has legs and if so, how many.) Include any eggs or other unusual life signs. Note the number and in which layer(s) organisms were found.
5. Place the organisms in a petri dish for further study with a hand lens or dissecting microscope.
6. If field guides are available, have one group member try to determine the type and name of each organism.
7. STOP & THINK Try to determine the roles or niches of the organisms you have found. For example, are the organisms plant- or animal-eaters? Record your reasons.
8. Go to a group which looked at a soil sample taken from a different location. Describe how the organisms in the two samples are similar and different, and speculate why this might be.

ANALYSIS

Write responses to the following in your notebook.

1. Describe the relationships among the plants, the animals, and the soil in the soil you examined.

2. What characteristics or factors did you use to determine the niche of each of the organisms?
3. What do you think might happen if two different populations of organisms with similar niches moved into the same habitat?
4. Describe the niches for the organisms in your bottle ecosystem.

What's for Lunch?

INTRODUCTION Many of the interactions in an ecosystem involve food. Food provides living things with the building blocks and the energy they need to carry out the processes of life. Think about what you have eaten today and trace it back to its original source. If you ate an egg, you know it came from a chicken which fed on corn, and the corn used sunlight, air, and water to make food. This is a simple example of a *food chain*, a linked feeding series. (Another is shown in Figure 2.1.) In natural ecosystems, food chains may be simple or complex. They include *producers* (autotrophs that make their own food by photosynthesis), *consumers* (heterotrophs that eat other organisms), and decomposers (organisms that break down the remains of other organisms into simpler substances).

In this activity, you will also be examining how individual food chains become interconnected: A consumer in one food chain often consumes organisms in another food chain, forming a food web. You will use descriptions of organisms found in a freshwater marsh ecosystem and connect them according to their feeding relationships and the transfer of energy. Each of the biota cards in this activity represents a *population,* a group of individuals of the same kind living in a particular area within a community of organisms.

snapping turtle
duckling
duckweed

Figure 2.1
A food chain.

MATERIALS NEEDED

For the class:

- 17 biota cards
- 1 cardboard "sun"
- 1 ball of yarn or string
- 1 scissors
- cellophane tape
- 1 meter stick or other pointer

▶ PROCEDURE

PART A: CREATING A FOOD WEB

1. Read through the entire Procedure before beginning this activity.
2. Read the information on your biota card.
   a. Place the sun and the producer cards on a large table or on an open floor space.
   b. Connect each producer to the sun with yarn or string and tape.
   c. Connect each consumer to the producer(s) on which it feeds, using yarn or string and tape.
   d. Continue the connections until all the biota cards have been linked.
3. Look at the arrangement. You have just created a marsh *food web*.
4. **STOP & THINK** Find all the food chains that make up this food web. What do they have in common? Record your response in your notebook.

NOTE: Some of the organisms eat (or are eaten by) more than one kind of organism, and will require yarn connections to each of the others. Therefore, depending on the overall arrangement of the cards, the lengths of the yarn pieces will vary. Also, as more organisms become connected to others with the yarn pieces, the yarn may become entangled. Be careful to keep these connecting links as straight and untangled as possible.

PART B: LOCAL EXTINCTION

An insecticide that kills mosquitoes and leaf beetles has been sprayed over this marsh ecosystem. What do you think will happen? The following procedure simulates what will happen in this community of organisms.

5. Locate and remove the mosquito card, trace the yarn pieces that connect to it and remove each piece of yarn along with the mosquito card. As these yarn links are removed, note the food source(s) of each organism that eats the mosquito.
6. Locate and remove the leaf beetle card, trace and remove all yarn pieces as in step 5.
7. Continue the simulation by looking at the remaining consumers. Do any consumers no longer have any food source in this ecosystem? If so, they also must be removed.
8. Continue the simulation until the full impact of the event is visible.
9. **STOP & THINK** Which organisms remain? Describe in your notebook how this event affected the food web as a whole and how it affected separate food chains. How might the ecosystem recover?

▶ ANALYSIS

Write responses to the following in your notebook.

1. What general principles can you state from this example? Give a specific example for each principle.

2. If the American black duck disappeared instead of the insects, what do you think would happen to the rest of the community? Explain your reasoning.
3. Where do decomposers fit into the food web? Why is their role necessary for life?

MANAGING MOSQUITOES

ECOLOGY: A NEW GREAT CHAIN OF BEING

(Excerpted from Ecology: A New Great Chain of Being, by Gordon Harrison, Natural History Magazine, *December 1968, pp. 8–17.)*

Ecology has been defined as the science of the interrelationships of creatures to each other and to their environment. There are many other ways of saying the same thing. I was talking recently to a biologist who had served for five years as a pest control officer in Borneo. He told me that some years ago the World Health Organization launched a mosquito control program in Borneo and sprayed large quantities of DDT, which had proved to be very effective in controlling the mosquito. But, shortly thereafter, the roofs of the natives' houses began to fall because they were being eaten by caterpillars, which, because of their particular habits, had not absorbed very much of the DDT themselves. A certain predatory wasp, however, which had been keeping the caterpillars under control, had been killed off in large numbers by the DDT. But the story doesn't end there, because they brought the spraying indoors to control houseflies. Up to that time, the control of houseflies was largely the job of a little lizard, the gecko, that inhabits houses. Well, the geckos continued their job of eating flies, now heavily dosed with DDT, and the geckos began to die. Then the geckos were eaten by house cats. The poor house cats at the end of this food chain had concentrated this material, and they began to die. And they died in such numbers that rats began to invade the houses and consume the food. But, more important, the rats were potential plague carriers. This situation became so alarming that they finally resorted to parachuting fresh cats into Borneo to try to restore the balance of populations that the people, trigger happy with the spray guns, had destroyed.

ANALYSIS

Write responses to the following in your notebook.

1. Create a flow chart that shows the result of the spraying of DDT.
2. What are some reasons this area was so seriously affected by the insecticide?

3. How might the government of Borneo have controlled the mosquito growth without such extreme consequences?
4. What conclusions can you draw about the importance of any one species in an ecosystem?

EXTENDING IDEAS

In early spring of 1963, a quiet aquatic biologist and science writer was interviewed in her Maine coastal home for a CBS television show. The program would be called "The Silent Spring of Rachel Carson." Before the show could be aired, three of the five original sponsors withdrew their support, and CBS received more than a thousand letters about the upcoming show, most criticizing it. When the program was broadcast as scheduled, it also included presentations by four United States government representatives and a prominent chemist of a chemical manufacturing company. What science topic could be that controversial?

Six months previously, Rachel Carson's book *Silent Spring* had been published. She had carefully and scientifically documented the destructiveness to the natural environment of chemical pesticides. Carson stated that their widespread use was potentially as damaging to life as nuclear explosions. She urged scientists and government officials to conduct further research in order to determine the effect of DDT and other pesticides on soil, water, animals and their food webs, as well as on human health and life.

Read *Silent Spring*—chapters one, six, and eight are particularly relevant to the concepts you are studying in this module—and write an essay that covers any or all of the following:

- the relevancy of Carson's ideas and concerns then and today
- whether her ideas have influenced current thinking and if so, how
- the place of pesticides in the modern world (developed and developing countries)
- what interested and/or surprised you in the book

Every June, approximately 2,200 amateur ornithologists and bird watchers across the United States and Canada join in an annual bird count called the breeding count survey. This survey, which began in 1966, has uncovered some interesting trends in bird populations. While birds such as robins, starlings, and blackbirds prosper around humans many other populations of colorful forest birds have severely declined. What is causing the devastating losses? Destruction of critical habitats is clearly a major issue, predation and parasitization also present a growing threat. Research the various ways in which humans and other organisms may be contributing to the decline in the bird populations as well as ways in which people are attempting to reverse the trend. You may also want to join some bird watchers in your area that participate in the annual breeding count survey and contribute to the research.

Owls, hawks, and eagles all belong to a group of birds called raptors or birds of prey. They actively hunt small vertebrates for food, particularly voles, shrews, and mice. Raptors consume bones, feathers, and hair when eating prey. After the bird has digested its meal, this undigestible material is rolled and compacted in the raptor's digestive tract to form a pellet, which is then passed through its body. Since owls tend to swallow their prey whole, their pellets are apt to contain whole skeletons.

Obtain an owl pellet from the field or local Audubon reservation. Wearing latex gloves, record its mass, length, and diameter. Place the pellet in a petri dish and carefully dissect it with teasing needles. Place the bones in the cover of the dish. Sort the pieces and assemble the skeleton. Try to identify the prey.

ON THE JOB

ZOOKEEPER The young Siberian tiger, Sheba, had just arrived at the zoo. As the zookeeper in charge of large cats (lions, tigers, and so on), Miguel will be overseeing the tiger's care during her stay. Sheba is on loan to the zoo in order to mate with a male Siberian tiger already on site. Because of the rarity of Siberian tigers today, Miguel's zoo is taking part in a global project to strengthen the gene pool of the Siberian tiger by mating females from some zoos with unrelated males from other zoos.

Nearly every child loves animals and waits expectantly for a visit to the zoo. Miguel spent many hours at the zoo as a child because his mother was a staff veterinarian at a large metropolitan zoo. She often brought Miguel to work with her and allowed him to tag along with different zoo workers, provided he not get in their way. He usually ended up helping Mr. Polk, the senior zookeeper, with his everyday tasks.

In high school, Miguel enjoyed his biology courses and made sure to take plenty of math and other science classes. He followed up in college with a major in zoology and a minor in communications, because he knew that a zookeeper is the first line of public relations for an animal park. The university he attended included a teaching zoo where students were able to earn credits for hands-on experience with animals. Not many schools offer this opportunity, but he specifically wanted as much hands-on work as possible.

The time Miguel will spend caring for Sheba will be only a small part of each busy day. He is in charge of the daily cleaning and maintenance of all the large-cat enclosures, and of the proper feeding of the animals under his care. As the keeper of many animals, and having other responsibilities, he does not have time to do all of the work himself, but he supervises other staff members and student interns as they complete some of these tasks. The animals' daily care requires Miguel to work weekends and holidays. Because of the hands-on aspect of his job, which can entail lifting heavy food bags and constructing different structures, he has to make sure he is physically fit.

When new animals arrive or a new exhibit is planned, Miguel helps to design and build any new enclosures, and to choose and care for the plants in and around the enclosures, making sure that the new homes closely resemble the animals' natural habitats. At the same time, he has to be primarily concerned for the safety of zoo visitors, being confident that no one can get into the enclosure and that warning signs are posted where needed.

Miguel also has the pleasure of introducing some of the animals under his care to the visiting public. A couple of times each week, he presents an animal showcase, bringing a few of the smaller animals onto a stage and explaining a little about them to the audience. For instance, he often brings the youngest members of his cat entourage onto the stage. The children gasp and the adults look on with interest. He describes the cub's life in the wild, what it eats, how it hunts, and then answers any questions the audience might have.

Miguel has to be extremely familiar with all the cats at the zoo so that he can detect any subtle changes in their behavior caused by illness or increased stress. Shortly after Sheba's arrival at the zoo, he noticed some changes in her behavior. And, after four months, Miguel was thrilled to witness the birth of her cub—named Tomorrow, because he was the first step toward keeping the Siberian tiger from extinction.

Round and Round They Go

PROLOGUE What makes Earth able to support life? In the last learning experience, you noted how populations of producers and consumers in an ecosystem are joined in food chains that create an interlinking food web. In this learning experience, you will examine the flow of energy through the *trophic* (or feeding) levels of a typical food chain. Then you will identify the essential resources needed for the preservation of life and see how these resources are cycled through the larger ecosystem that is planet Earth. These *biogeochemical cycles* transfer materials (organic and inorganic) among organisms, from organisms to the abiotic environment, and back to the organisms, thus sustaining life on Earth.

The Flow of Energy

INTRODUCTION Think about the number of blades of grass it takes to feed a prairie dog, and the number of prairie dogs to feed a coyote. Why is there such a difference? Why are there fewer consumers in each successive trophic or feeding level?

All of life's processes require energy. After a consumer obtains nutrients from its food source, it processes the nutrients to obtain energy. The energy is then used by the organism for growth, movement, and a variety of other activities.

Does the amount of energy available in a food source depend on its trophic level? In the process of *photosynthesis,* producers convert light energy (from the sun) to chemical energy (in the form of sugar) in their chlorophyll-containing structures. *Herbivores* (plant-eaters) are consumers that use the chemical energy harnessed by producers as their food source; *carnivores* consume herbivores. In this activity, you will investigate the flow of energy through organisms in successive trophic levels of a food chain.

Table 3.1
Energy loss along a food chain.

TROPHIC LEVEL	ENERGY SOURCE	% ENERGY LOST DURING METABOLISM	% ENERGY LOST TO WASTE
Primary producer*	Sunlight energy	60%	20%
Herbivore**	Primary producers	65%	20%
Primary carnivore***	Herbivores	70%	25%
Secondary carnivore ****	Primary carnivores	70%	25%

*Primary producer—an organism that synthesizes food molecules from inorganic compounds by using an external energy source.
**Herbivore—an organism that eats only plants.
***Primary carnivore—an organism that eats an herbivore.
****Secondary carnivore—an organism that eats a primary carnivore.

TASK

1. Sunlight provides energy to the primary producer—20,000 kilocalories (kcal) per square meter. (A *calorie* is the unit of measurement for the energy produced by anything when burned or oxidized; a kilocalorie is equal to 1000 calories. Foods oxidized in the body produce so much energy that when we speak of the calories of energy in foods, we are actually referring to kcal.)

   The producer is consumed by the primary consumer, but some of the energy is used during metabolic activities of the producer, or lost as heat. Using the percentages from Table 3.1, calculate the kilocalories used by each trophic level and the kilocalories available to be used in the next trophic level.
2. Construct a line graph that shows the energy transfer in this ecosystem. Make sure to represent the energy in a way that makes obvious the amount of energy being transferred.
3. Write responses to the Analysis in your notebook.

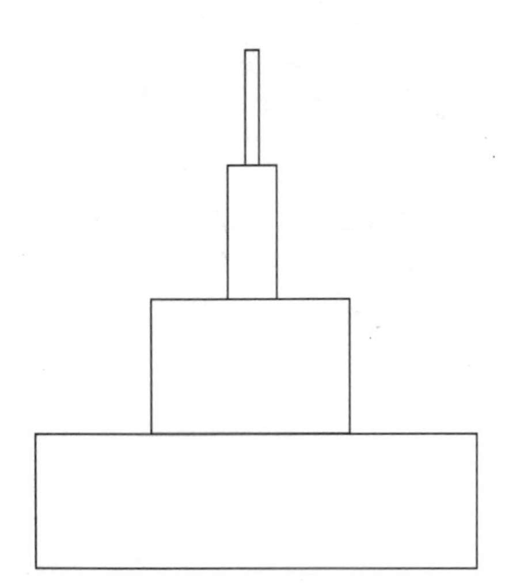

Figure 3.2
Pyramid of numbers.

ANALYSIS

1. How much of the original 20,000 kcal of energy is available for the secondary carnivore? How much energy would be available to a tertiary carnivore?
2. Explain the flow of energy through trophic levels.
3. Is there more energy transferred to the consumer after eating a pound of rice or a pound of meat? Explain your response.
4. Explain the concept that is illustrated by the pyramid in Figure 3.2.
5. Observe Figure 3.3. How many trophic levels do you see in this web? Designate a name for each level and give an example from the web.

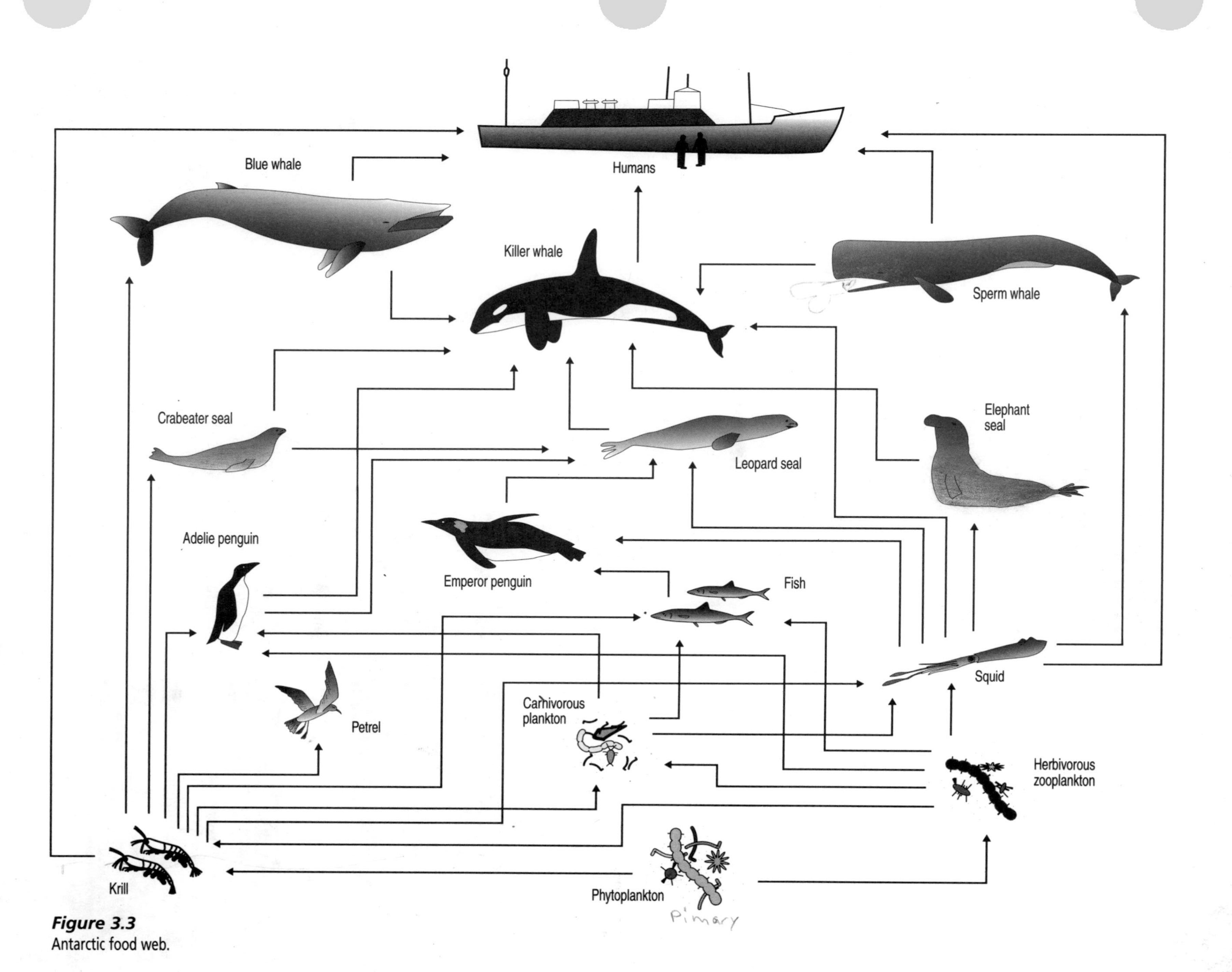

Figure 3.3
Antarctic food web.

It's Elemental

INTRODUCTION What resources do plants and animals need to maintain life? In this investigation, you will use snails and a simple freshwater plant known as *Elodea* to explore the interrelationships among animals and plants and the resources found in the environment. You will use the chemical indicator bromothymol blue, which turns green or yellow in the presence of an acid, to trace the presence or absence of carbon dioxide. When dissolved in water, carbon dioxide forms a weak acid, carbonic acid.

MATERIALS NEEDED

For each group of four students:

- 4 pairs of safety goggles
- 4 culture tubes or test tubes with tight-fitting covers or stoppers
- 1 test tube rack
- 1 wax marking pencil
- 100–200 mL distilled water (or "aged" tap water)
- 1 eyedropper
- 1–2 mL of 0.1% bromothymol blue solution (aqueous)
- 4 small water (pond) snails
- 2 sprigs of *Elodea*
- access to a fluorescent lamp

SAFETY NOTE: *Always wear safety goggles when conducting experiments. After handling snails or other animals, wash your hands.*

PROCEDURE

1. Obtain four test tubes and place them in a test tube rack. With a wax marking pencil, label the tubes 1, 2, 3, and 4; then pour distilled water into each of the tubes until they are 3/4 full.
2. Using an eyedropper, add four drops of bromothymol blue solution to each tube.
3. Tightly seal tube 1 with a stopper and return it to the test tube rack (see Figure 3.4 on the next page).
4. Tilt tube 2 slightly and gently add two snails. Place the stopper in the tube and place it in the rack.
5. Add one sprig of *Elodea* to tube 3. Stopper the tube and place it in the rack.
6. Add two small snails and one sprig of *Elodea* to tube 4. Place the stopper on the tube and place the tube in the rack.
7. **STOP & THINK** Write a prediction in your notebook of what you think will happen in each tube.

CAUTION: *Use caution when handling chemicals. Bromothymol blue is a skin and eye irritant. Avoid contact with skin and clothing. If contact occurs, flush the affected area with water immediately.*

8. **STOP & THINK** What is the purpose of tube 1? Write your response in your notebook.

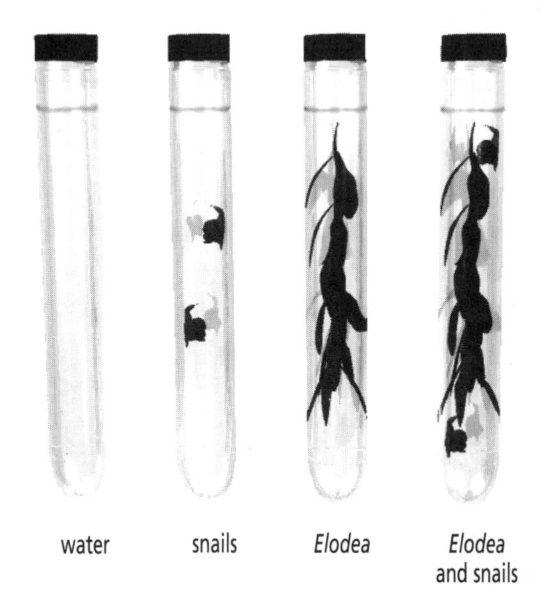

Figure 3.4
Closed living systems.

9. Set the test tube rack under a fluorescent lamp.
10. Observe the tubes daily for one to two weeks, each time noting the color of the water and the apparent health of the organisms. Record in a chart in your notebook.
11. At the end of the investigation, dispose of any dead organisms. If possible, place your live organisms in an existing aquarium; or, prepare an aquarium tank for them. Snails and *Elodea* are native throughout the United States, and may be placed in any pond or stream.

► ANALYSIS

At the completion of the investigation, write responses to the following in your notebook.

1. In which tube(s) did you note a color change? What caused the change in each tube?
2. Which organisms remained healthy? Explain why you think they did.
3. What gases were cycling in tubes 2–4? How do you know?
4. Based on your knowledge from this experiment, draw a picture of the cycles involved.
5. What do you think would happen in each tube if they were placed in the dark?
6. In what ways do the test tubes resemble the ecocolumns? How do they differ?

An Infinite Loop

INTRODUCTION Organic nutrients such as carbohydrates, proteins, and lipids are made up of elements including carbon, oxygen, hydrogen, and nitrogen. However, these elements are also present in the environment in inorganic forms.

The amounts of carbon, oxygen, and nitrogen in the present atmosphere have remained nearly the same since life came into existence about 3.8 billion years ago. That means that the oxygen you breathe could also have been inhaled by your great-grandparents or by George Washington. And the carbon in the food you ate for dinner might once have been part of a dinosaur! How is this possible?

In this activity, you will be assigned to follow the flow of one of the following cycles: carbon–oxygen, nitrogen, or water. You will read information about that cycle and discuss it with a partner to clarify your understanding. Then you will diagram the cycle.

MATERIALS NEEDED

For each group of three students:

- 3 sheets of unlined white paper (11 x 14 inches, legal)
- felt-tip markers or colored pencils

TASK

1. Separate from your group of three and join other students who have been assigned to the same cycle. With a partner from this new group, read the information given, and discuss how to draw this cycle.
2. Create a labeled diagram of the cycle on a sheet of paper (11 x 14 inches). Include all applicable biotic and abiotic factors and arrows to show the cycling.
3. Return to your group of three and explain the cycle to them.

NOTE: Each person needs to create a diagram.

CARBON-OXYGEN CYCLE

1. These two linked cycles provide plants and animals with energy and the materials for the basic building blocks of life. Carbon and oxygen have independent cycles, but often travel together. Carbon, oxygen, nitrogen, and hydrogen are bonded together and form biomolecules—that is, carbohydrates, proteins, and lipids. Carbon (C) is found in all living things in biomolecules. It is also found in the atmosphere, as carbon dioxide (CO_2). Oxygen (O_2) makes up 21% of Earth's atmosphere and is found dissolved in fresh and ocean waters.

2. In the process of photosynthesis, plants take in water (H_2O) from the soil and carbon dioxide from the air. They are converted into biomolecules such as sugars or carbohydrates. The byproduct—oxygen—is released into the air.
3. In the process of respiration, all living things (including plants) use oxygen to "burn" sugars, producing water and releasing carbon dioxide as a byproduct into the air.
4. Animals eat plants or other animals and break down the consumed organism's complex biomolecules into simpler biomolecules, releasing energy that is now available for the consumer.
5. As land animals and plants die, decomposers use oxygen from the air or soil to break down the dead organisms' carbon-containing biomolecules, and carbon dioxide is released into the atmosphere.
6. Over long periods of time, pressure from overlying soil or water can compress carbon from dead organisms into peat, and then into fossil fuel such as coal, natural gas or oil. When humans burn fossil fuels, the carbon in the fuel combines with oxygen in the air and releases carbon dioxide into the air.
7. Photosynthetic phytoplankton in the ocean take carbon dioxide from the air or ocean water and incorporate the carbon into biomolecules, releasing oxygen.
8. As aquatic animals and plants die, decomposers oxidize the dead organism's carbon compounds, releasing carbon dioxide into the ocean water. Organic material also sinks to the aquatic floor where, over time, the carbon can be compressed into fossil fuels.

NITROGEN CYCLE

1. Nitrogen gas (N_2) makes up 78% of Earth's atmosphere. Plants and animals can not use nitrogen in this gaseous form. Nitrogen is also found in ocean water and in nitrogenous compounds located in soil.
2. *Nitrogen-fixing* bacteria in the soil, and cyanobacteria in the ocean, are able to convert nitrogen gas into ammonium (NH_4) or ammonia (NH_3). Ammonia in soil also comes from excreted animal waste.
3. Other *nitrification* bacteria take the ammonia in the soil and combine it with oxygen from the air or soil, forming nitrites (NO_2) and then nitrates (NO_3). Plants convert these nitrogen-containing compounds into biomolecules such as proteins, amino acids, or nucleic acids. Some nitrates in the soil are leached into groundwater and eventually carried into waterways leading to the ocean.
4. Both on land and in the ocean, consumers eat other animals or plants and use the nitrogen in the protein of these organisms. The nitrogen is incorporated into other proteins the consumer can use, or the proteins are broken down to produce energy; the nitrogen combines with hydrogen to form ammonia and excreted as waste.

5. Decomposers in the soil and in the ocean break down the nitrogen-containing biomolecules in dead organisms and release nitrogen gas into the air and water. This process is known as *denitrification.*
6. As aquatic plants and animals excrete waste or die, this organic material sinks to the ocean floor. Some bacteria that decompose nitrogen compounds convert nitrates to nitrites; others are able to convert nitrates back to molecular nitrogen (N_2), thus completing the (aquatic) cycle. Seasonal ocean currents bring water and nutrients back to the surface of the ocean (a process called upwelling), where nitrogen gas is released into the air or converted into ammonia by bacteria.
7. Other types of bacteria in the ocean do the following: convert nitrogen gas into ammonia; convert ammonia into nitrites; or convert nitrites into nitrates.

WATER CYCLE

1. The oceans are the reservoir for 97% of the water on Earth.
2. The heat from the sun is the force that drives the water cycle. The sun heats the ocean water, which evaporates into the air. Salt remains in the ocean. The sun also causes evaporation from lakes and rivers.
3. The evaporated water in the air condenses into clouds which, when cooled, release rain, snow, and other forms of *precipitation* back into the ocean and onto land.
4. Water is released into the air from leaf surfaces during photosynthesis. During respiration, all living things release water vapor into the air.
5. When precipitation falls on land areas with vegetation, the plant roots hold the water in the soil. This water (called groundwater) is available for use by plants and animals. When precipitation falls on land areas without vegetation, excess water can wash away topsoil into lakes or rivers.
6. Precipitation may also fall on lakes, rivers or the ocean.
7. Water in lakes and rivers may eventually return to the ocean.
8. Water that does not flow into lakes or rivers, and is not used by plants and animals, moves through the ground. Large underground reservoirs (aquifers) are located beneath the earth's surface. This water slowly makes its way back to the ocean.

▶ ANALYSIS

Write responses to the following in your notebook.

1. Diagram, in words and/or pictures, the nitrogen and water cycles. Include all biotic and abiotic factors.
2. Draw a carbon cycle, using a black marker or crayon. On the same sheet of paper, draw an oxygen cycle, using a blue marker or crayon.
3. Choose one major biotic or abiotic factor in each of your cycles and describe the consequences if this factor were not present. Explain. your response.
4. Kamo no Chomei, a Japanese author, wrote, "The flow of the river is ceaseless and its water is never the same." Using your knowledge of the water cycle, explain this quotation.
5. All of the elements in these cycles are finite (of limited amounts). How is this fact important in your thinking about the cycles?

Figure 3.5

Overlap Among the Cycles

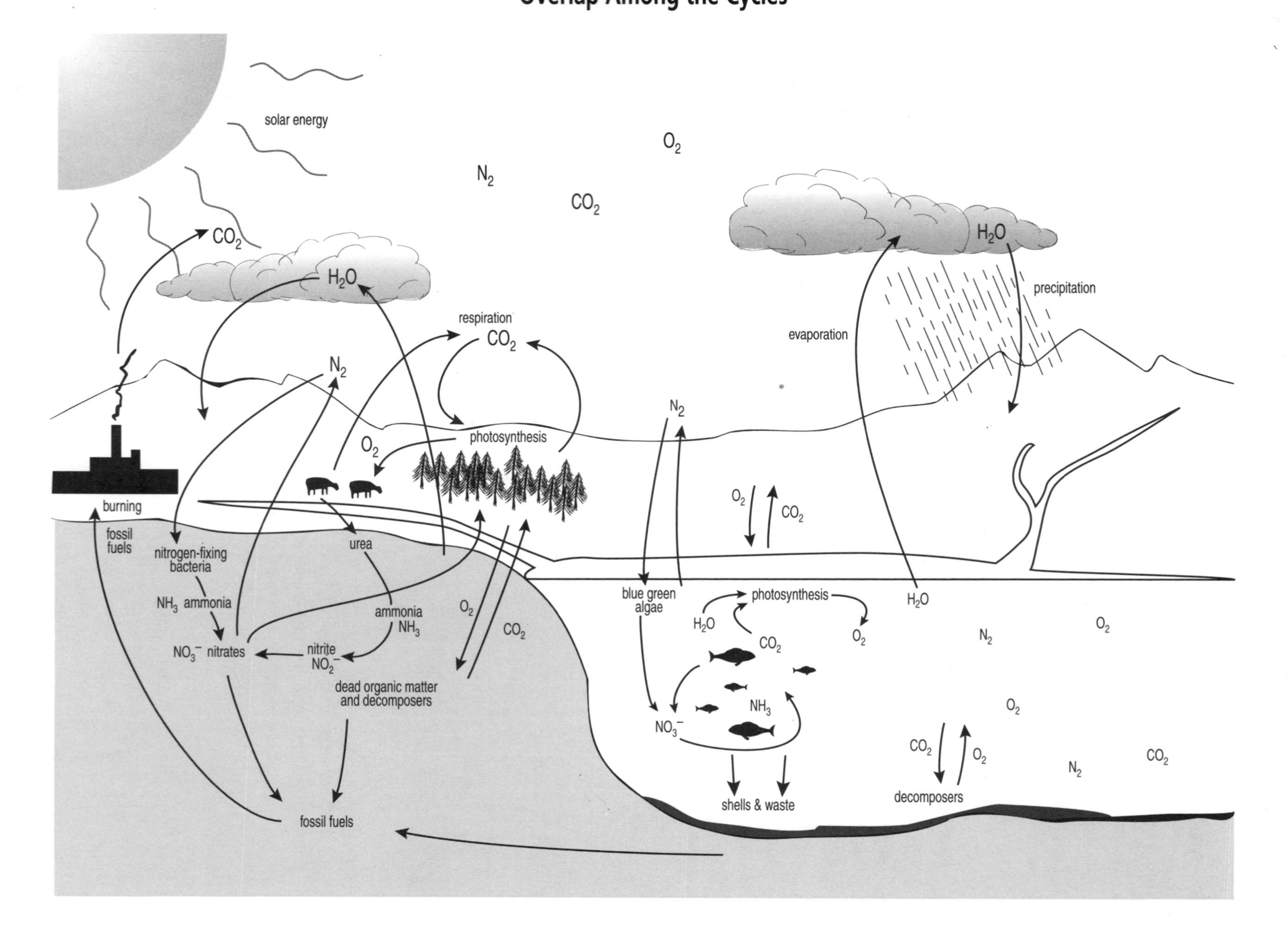

EXTENDING IDEAS

- There are approximately 36 chemical elements which cycle through the biosphere. Research one or two others (such as phosphorus, sulfur, or calcium) to see how their cycles complement the cycles you already explored.
- Conduct a soil analysis of your square meter and test for the presence of water or nitrogen.
- The burning of fossil fuels—coal, oil, natural gas—is constantly adding carbon dioxide into the air. As a result of excess carbon dioxide, a greenhouse effect on Earth is occurring. The carbon dioxide in the air traps the sun's energy, not allowing the heat to escape. Determine why the greenhouse gases might be a cause for worry. Research the current thinking about the link between burning fossil fuels and global warming.
- Each year the streams and rivers of the United States carry away 4 billion metric tons of sediment. Wind blows away another 1 billion metric tons. This loss of topsoil reduces soil fertility and crop production. Once it is lost, soil cannot easily or quickly be replaced. Soil erosion has become a serious environmental and economic problem. Research some of the methods currently being used for reducing soil erosion.

ON THE JOB

METEOROLOGIST "There is a hurricane warning in effect for coastal Alabama, Mississippi, and Louisiana. Hurricane Jasmine is traveling northwest at 35 mph and carrying sustained winds in excess of 120 mph. She is expected to make landfall just before 8 o'clock tonight. Residents should be prepared for torrential downpours and possible hail. . ."

Melinda turned off the microphone. A meteorologist with the Mobile, Alabama branch of the National Weather Service, she had been at work for hours. She was tracking what might be the strongest, most dangerous hurricane to hit the northern Gulf coast in decades. Surrounded by all of the latest weather forecasting equipment, Melinda was watching closely for any changes. Although severe weather meant long hours of work, it always renewed her love and respect for the forces of nature.

Melinda grew up in the Midwest, where hurricanes were never a problem but tornadoes were much feared. She saw the humbling destructiveness of a tornado just once. As much as the ferocity of storms excited her, the calmness of a clear day with white puffy clouds gliding overhead was just as intriguing. It was this intrigue that lead her into the field of meteorology.

While in high school, Melinda researched meteorology and learned that most jobs in that field were with the National Weather Service (NWS), a branch of the National Oceanic and Atmospheric Administration (NOAA). Melinda chose a college that offered a degree program in the field. Had she been unable to attend a college offering a meteorology degree, she could have prepared to get a job in the field by taking college courses required by the NWS. Those courses include meteorology-specific classes such as weather analysis, forecasting, and dynamic meteorology; and general courses that include calculus, physics, computer science, and statistics. Melinda also discovered that forecasting the weather was not her only option. Some meteorologists spend their days doing research on the physical and chemical properties of the atmosphere and on factors affecting the formation of clouds, rain, snow, and hail. Others study past records of long-term weather patterns in specific regions in order to advise workers about building design and agriculture.

But Melinda was most excited about the prospect of forecasting the weather. After receiving a bachelor's degree, she landed a job with the NWS in Mobile. Her entry-level position entailed collecting atmospheric data from weather balloons and from other technological data sources such as Doppler radar, which can detect rotational patterns in violent storm systems. She was asked to analyze the data collected to aid the other forecasters. After spending a little over a year in that position, she started her climb toward being the primary meteorologist in the Mobile branch. Now she is doing what she loves.

Hours later, Melinda picked up the microphone. "The skies are clear over Mobile, and the now-tropical storm Jasmine continues to weaken as she travels inland. There have been few storm-related injuries and no deaths. The next few days are expected to be calm, with sunny skies and warm temperatures."

Population Pressures

PROLOGUE What factors influence population size? Although populations may generally stay about the same size from year to year, changes in environmental conditions or in biotic factors may cause a resultant change in population size. Among prairie meadow mice, for example, a cold summer may result in an increase in reproduction. As the number of mice increases, their *predators,* such as the hawks and coyotes who feed on mice, will also start to increase. If normal weather patterns return in subsequent years, reproduction of the mice will decline. The hawk and coyote populations will begin to decline in turn as their *prey* decrease in number, restoring the system to equilibrium over time. However, if cool summers persist for long periods of time the populations might shift to a new equilibrium.

When ecologists examine the populations of organisms, they look for trends in how populations increase and decrease over time. They then use this information to analyze the dynamics of the populations and the entire ecosystem.

In this learning experience, you will begin to examine the different patterns of population growth and determine the significance of these patterns over time. You will evaluate how the factors that influence the growth or reduction of populations interact, thus determining the population size at any given time. Finally, you will look at populations of organisms in northeastern forests and see how many interacting factors influence both Lyme tick-borne disease epidemics and outbreaks of gypsy moths.

UNSUPERVISED

INTRODUCTION If a population could live under ideal conditions—having all the food, space and other resources they need, no competition for resources, and no predators—such a population would show its *biotic potential*. Existing organisms would reproduce to their fullest, as would subsequent generations. Such a large increase in numbers is called *exponential growth* and when placed on a graph it produces a *J-shaped curve* (see Figure 4.1).

However, conditions in nature are seldom ideal. Predator–prey relationships are the norm, and a variety of other environmental factors hinder such unlimited growth. In this activity, you will examine one population, that of yeast, growing undisturbed in a sugar culture for five days. What will happen to this population over time? What factors will influence changes in the yeast population?

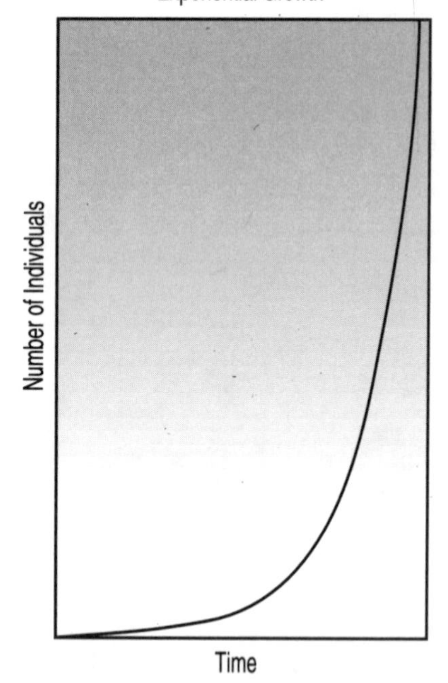

Figure 4.1
The J-shaped curve shows the exponential growth of a population under ideal conditions.

MATERIALS NEEDED

For each pair of students:

- 1 sheet of graph paper

PROCEDURE

1. Examine Figure 4.2 on page 37. It represents samples taken over time from a yeast population growing in a nutrient-rich culture flask. The samples were placed on a glass slide with a grid (to make counting easier), and viewed through a microscope.
2. **STOP & THINK** What type of growth curve do you predict will occur?
3. Create a data chart in your notebook in which you will record the number of yeast cells in each of the samples in areas A, B, and C, and the time intervals. Insert a fourth column for the average number of yeast cells in all areas.
4. Count the yeast cells (represented as small circles) on each slide.
5. Record your data in the chart you created. Calculate the average of the three areas to one decimal point. Check your numbers with your partner.
6. Construct a graph of your data. (Only about 1/1000 of the original yeast population was placed on the counting slides. This sampling limitation should be noted on your graph.)

Figure 4.2

Yeast populations grown in nutrient rich culture.

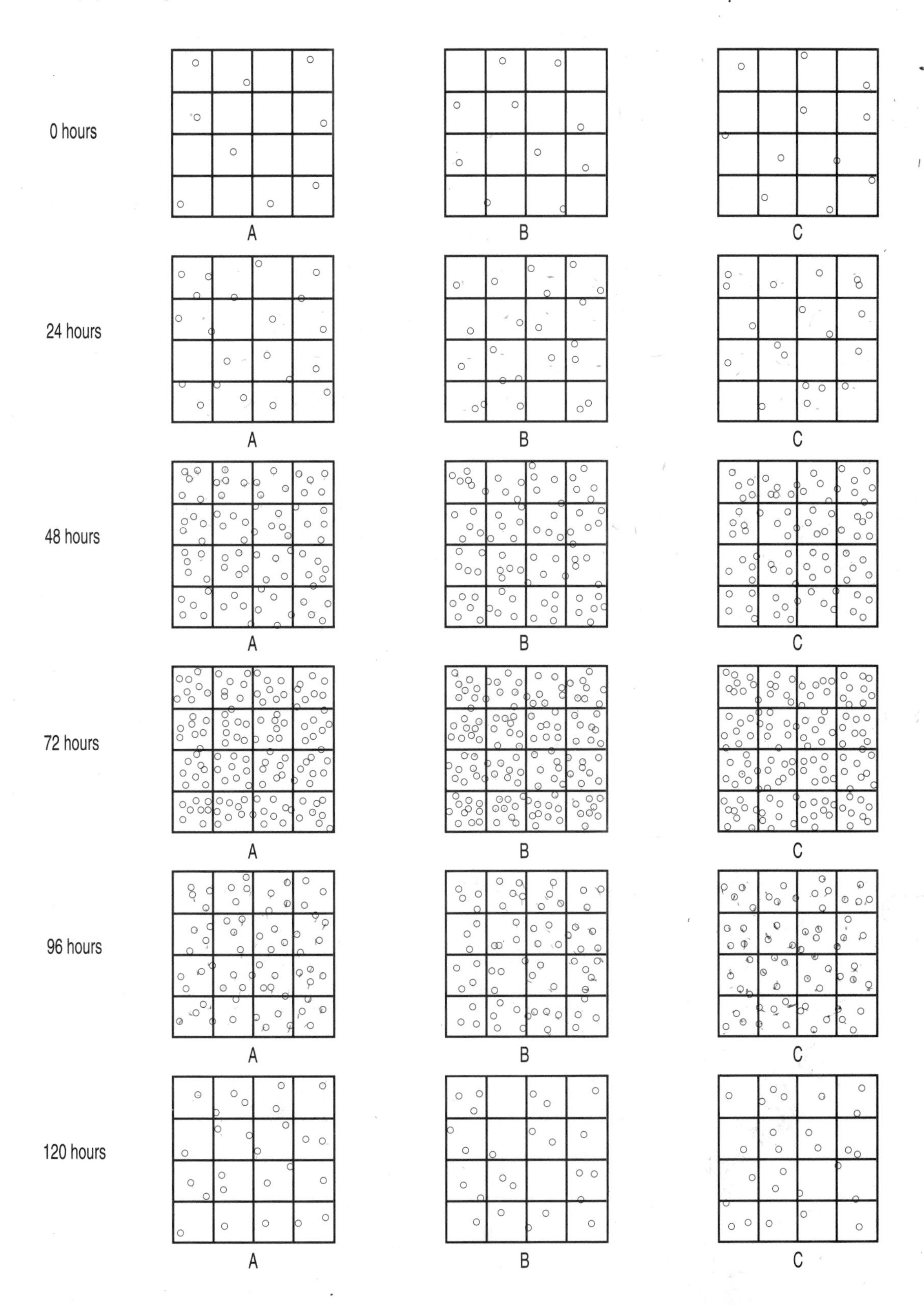

ANALYSIS

Write responses to the following in your notebook.

1. During which time intervals was the population growth most rapid? Why?
2. What was happening during the other time intervals?
3. Use Figure 4.3 below to identify and label the stages of growth on your graph.
4. If the yeast were started in a larger flask, would the graph be different? In what way? Why do you think so?
5. How is the yeast growth curve similar to and different from an exponential growth curve? What might be some reasons for these differences?
6. The *carrying capacity* is the maximum number of organisms an area can hold; the population remains at a steady rate for a period of time (see Figure 4.3). What might cause the carrying capacity of a population to change?

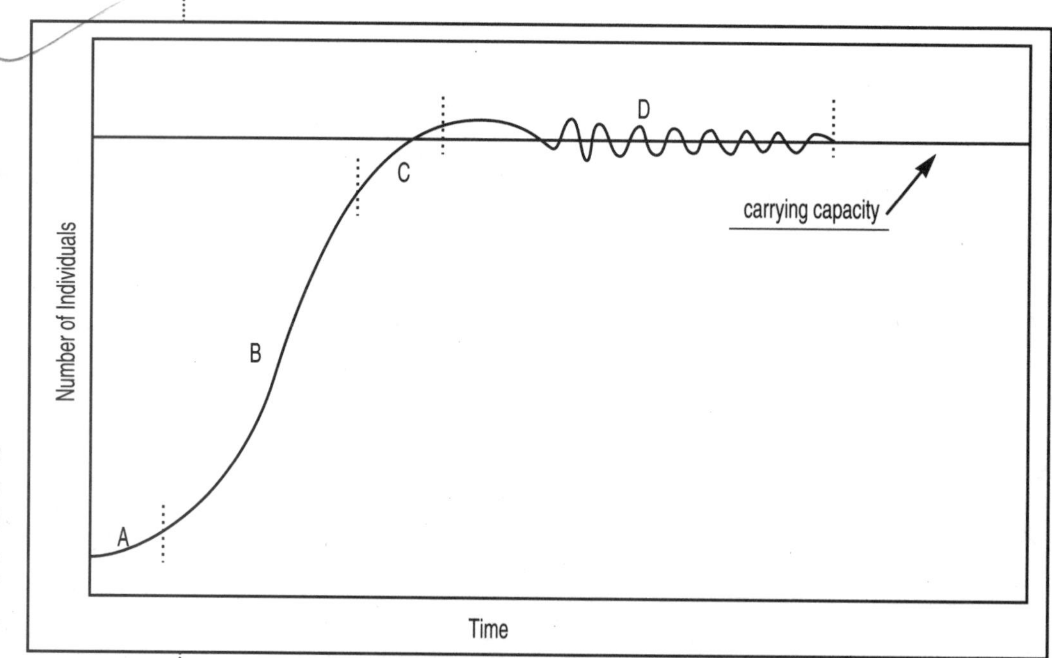

Figure 4.3
Growth curve stages: (a) slow growth, (b) exponential growth, (c) slowing growth, (d) growth leveling off near the carrying capacity.

THE LYNX AND THE HARE

INTRODUCTION Is the "boom and crash" pattern you saw in the yeast culture typical of all populations? Remember that no animal, plant, nor insect exists in total isolation in the "real" world. Even within a small community there are hundreds, even thousands, of different populations and each of them has positive and negative effects on the others.

A variety of biotic and abiotic factors both stimulate and reduce population growth. For example, predators, disease, parasites, and competition with other organisms are biotic factors that reduce population size. Abiotic factors such as weather, lack of food, water, or shelter are also known to reduce population size. These *limiting factors* generally balance reproductive rates and keep natural populations below their biotic potential. What is the relationship between these limiting factors and growth factors?

In this activity, you will explore the relationship between populations of lynx and their prey, snowshoe hares. In nature, both predator and prey have evolved structures and behaviors to counterbalance each other and thus they coexist in a moderately stable relationship. In the following simulation, you will determine what happens to each population over many years.

MATERIALS NEEDED

For each group of four students:

- 100 hare cards (5 cm x 5 cm)
- 25 lynx cards (10 cm x 10 cm)
- 1 ruler or tape measure
- masking tape
- 1 sheet of graph paper

Figure 4.4
Lynx and hare.

PROCEDURE

1. Use masking tape to outline a 50 cm x 50 cm square on a flat surface to simulate an ecosystem.
2. Create a three-column table in which to record the following: round number (each round equals one year), the number of hares, and the number of lynx. You will begin with 6 hares and 1 lynx.
3. Take six hare cards and scatter them randomly within the ecosystem.
4. Take one lynx card and drop it from a height of 10–15 cm above the hares. Try to drop the lynx onto as many hares as possible.
5. Remove all of the hares caught by the lynx. A hare has been caught if the lynx square touches it in any way.
6. If the lynx catches three or more hares, it will survive and reproduce one kit (baby). When this happens get another lynx card. If the lynx catches fewer than three hares, the lynx will die. In this case, remove a lynx card. If the number of lynx in the ecosystem falls to zero, one new lynx immigrates into the area for the next round.
7. Each surviving hare reproduces one baby. To simulate this, scatter one new hare card in the ecosystem for each surviving hare. Six is

NOTE: You may wish to assign tasks within the group such as dropping the lynx card(s), removing caught (dead) hare cards, reproducing (or replacing by immigration) the hare and the lynx populations, and recording the new population for the hares and the lynx for the next round.

the smallest number of hares that will ever be in your ecosystem. If the number of hares in the ecosystem falls below six, then more hares immigrate into the area for the next round ensuring the minimum of six hares in the population.

8. This is the end of round one. Record the numbers of hares and lynx living in the ecosystem.
9. Use the new number of lynx cards and repeat steps 4–8 until you have completed 25 rounds.
10. Create a graph that illustrates the change in the lynx and hare populations over time (the rounds).

ANALYSIS

At the completion of the activity, write responses to the following in your notebook.

1. Describe your graph in words.
2. How is the population of hares dependent upon the population of lynx?
3. How is the population of lynx dependent upon the population of hares?
4. What would happen if all of the hares were removed from the ecosystem? Explain your response.
5. What do you think would happen if all of the lynx were removed? Explain your response.
6. Would you consider the lynx and the hare populations to be balanced? Explain the reasons for your decision.

A Friendly Warning

Ticks and Moths, Not Just Oaks Linked to Acorns

MICE THRIVE ON ACORNS AND DEER TICK LARVAE THRIVE ON MICE.

by Les Line. The New York Times, *April 16, 1996.*

A tangled cycle of events in northeastern forests that gives a reason for both Lyme disease epidemics and outbreaks of gypsy moths has been unraveled by ecologists. They have traced both events to the bumper crops of acorns that are produced every three or four years and to the white-footed mice that feed on them.

The bumper acorn crops influence the life cycles of mice and deer as well as the gypsy moths and the spirochetes that cause Lyme disease. Based on their theory, the two ecologists, Dr. Richard S. Ostfeld and Dr. Clive G. Jones of the Institute of Ecosystem Studies in Millbrook, N.Y., predict that the gypsy moth population in the Northeast will undergo one of its periodic explosions beginning this year. They expect it to build to a major defoliation in 1999 that will rival the devastating caterpillar blight of 1979–81 unless one of the moth's natural enemies, like a fungus or parasite, intervenes.

The two ecologists also warn that there will be tremendous numbers of deer tick nymphs, the vectors of Lyme disease, in the oak-dominated woodlands of New York and New England this spring and summer.

What the two ecologists call "the acorn connection" is laid out in the May issue of the journal *Bioscience* in an article written by them and Dr. Jerry O. Wolff, a biologist at Oregon State University. The study's principal findings come from long-term studies of gypsy moths and white-footed mice at the Mary Flagler Cary Arboretum, site of the Institute of Ecosystem Studies, and a 14-year study by Dr. Wolff of mouse populations at the Mountain Lake Biological Station in southwestern Virginia.

The forests at both sites are dominated by oaks, which have evolved a cunning strategy for reproduction. Their acorns are rich in the proteins and fats that give them a head start over other tree seeds, but that also make them a favorite food of mice and deer. So instead of producing large crops each year, which would foster a steady population of acorn consumers, the oaks have evolved a feast-and-famine regimen: occasional bumper crops, separated by years in which the consumers starve.

"The evolutionary response of oak trees appears to have been to produce such large crops of acorns that the various forest consumers are simply unable to eat them all, and some survive to become seedlings," the scientists concluded.

"Trees aren't as stupid as they look," Dr. Ostfeld said.

In the Virginia study, 1980, 1985, 1988 and 1989 were years of peak acorn production, while bumper crops

Continued on next page

were produced at Cary Arboretum in 1991 and 1994. By regular trapping, the researchers found that white-footed mice became most numerous in the summer after a bumper acorn crop but that their numbers dropped sharply the following winter or spring. Except in bumper years, mice run out of stored acorns by January.

the microorganism, called a spirochete, that causes Lyme disease. The tick larvae take a single blood meal from the mice, then molt a few months later into nymphs. The following spring, these Lyme disease-carrying nymphs attach themselves to mice and other mammals, including humans. They molt to become adults in late summer; the adult ticks wait in ambush for a deer to brush past, and the cycle begins anew.

"Without mice, the deer ticks would be a nuisance," Dr. Ostfeld said, "but they would not cause illness because they hatch from eggs free of the Lyme disease agent."

By measuring the number of ticks on the forest floor, the ecologists have shown that the larvae are most numerous in oak-dominated woods in the year following a bumper acorn crop; they peak in maple-dominated woods in years after poor acorn crops. That confirms the idea that deer are the ticks' principal transport system. "There is no evidence whatsoever that other factors, particularly weather patterns, influence tick abundance," Dr. Ostfeld said.

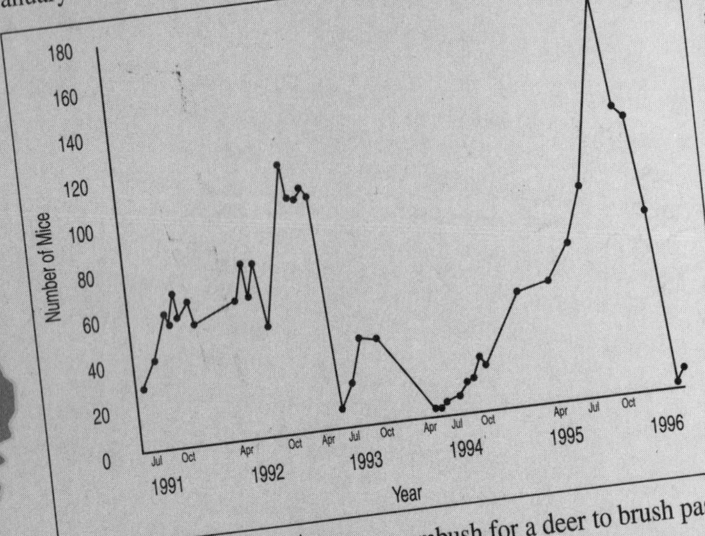

Changes in mouse population size.

The next actors in the drama are the white-tailed deer, which move into the oak-dominated areas of the forest during years of bumper acorn crops. The deer also import a cargo of adult ticks, which drop off and lay eggs in the leaf litter beneath the oak trees. The tick larvae, each the size of a pinhead, hatch the next summer, just as the woods are overrun with a peak population of their preferred host, the white-footed mouse.

It is from the mice, not the deer, that the tick larvae pick up

The spirochetes' special host, the white-footed mouse, is an appealing woodland rodent with deer-colored fur, big eyes and big ears; from [nose] to the tip of their long tail the mice measure six to eight inches. Besides their proclivity for acorns, white-footed mice have another habit that influences the forest: a taste for gypsy moths' pupae.

Gypsy moth caterpillars feed on oak leaves in the spring and early summer, then make cocoons and enter a pupal stage for about two weeks. The inch-long pupae, Dr. Jones said, are "tender, tasty morsels" for the mice. As long as the mice are moderately abundant, they will eliminate virtually all of the pupae, keeping gypsy moth populations firmly in check.

But when the mouse numbers suffer their inevitable crashes, in years of sparse acorn crops, a great many pupae survive and the moth population builds over the next several years, reaching a peak when their hairy caterpillars denude entire hillsides and damage or kill thousands of trees. At this point, Dr. Jones said, "you can't get enough mice to control the moths."

The wide-scale defoliation caused by gypsy moth outbreaks has many influences on the forest and its creatures. A major impact is on the oaks. Defoliation may delay or even eliminate the years of bumper acorn crops. The moths may also reshuffle the species composition of the forest by killing the oaks, encouraging the growth of saplings from other tree species. "The oak forest system under study is exceedingly complex," the ecologists said.

Like the soar-and-sink mouse populations, the number of gypsy moths is also prone to sudden collapse. After caterpillars defoliate many trees, the next caterpillar generation finds that food is short and that their natural insect and fungus parasites abound. The number of moths dwindles until conditions are right for their next surge.

To test the idea that white-footed mice keep gypsy moth populations in check, researchers at Millbrook counted the numbers of each in successive years. In the summer of 1993, the mouse population was moderately high and consumed almost all the gypsy moth pupae within eight days. But few mice were around the next year, so the pupae could survive uneaten long enough to hatch into moths.

Dr. Robert Colwell, professor of ecology and evolutionary biology at the University of Connecticut in Storrs, said: "This study is a stunning example of the importance to both human and ecosystem health of basic research in ecology and natural history."

Does this insight into the forest's natural mechanisms suggest ways to reduce the incidence of Lyme disease or the severity of gypsy moth infestations? So far, it seems not. Supplying extra feed to the mice in years of poor acorn crops might minimize gypsy moth outbreaks, the ecologists said, but the risk of tick bites would soar. Lyme disease, they said, has already resulted in "fear and distrust of nature" and has dampened the enthusiasm of many people for enjoying the forests.

ANALYSIS

Imagine that you are an ecologist who has been asked by the Fish and Wildlife Division of Virginia to create poster that will be posted in the Virginia forests. The poster's purpose is to inform hikers and other visitors of the danger years for contracting Lyme disease and for infestation of gypsy moths over the 10 years from 1997-2006. Use the illustration "Acorn Rollercoaster" as a guide for taking notes on the article. The poster must include the following:

a) A line graph indicating what happens to *each* population of organisms in each year. The graph should have the years along the x-axis and the approximate number of organisms on the y-axis. This graph requires you to predict what will happen based on the information in the article.

b) All five organisms—mice, ticks, deer, moths, acorns—must be plotted on the same graph. (Be sure to take into consideration the number of a specific type of organism generally found in a population. For example, there are generally a lot more insects in an ecosystem than in there are mammals.)

c) The graph should be annotated indicating years when their is a danger for contracting Lyme disease as well as years when their is a danger for gypsy moth infestation.

d) There should be a short paragraph to inform the hiker about how these predictions were made.

Your poster must be big enough and neat enough to be put on a wall and comfortably read from a distance of five feet, and should include all the information listed above. Accuracy of predictions, quality of visual presentation, and completeness will be key in how the information on the poster will be assessed.

EXTENDING IDEAS

- How can orbiting satellites, which enable people to forecast the weather make phone calls from cars, and choose from hundreds of television channels, be used to predict outbreaks of deadly disease? In an area of research called landscape epidemiology, scientists use features of a landscape to identify where and when infectious diseases might occur. This approach has been used to identify areas at risk for outbreaks of Lyme disease. By analyzing vegetation patterns and examining animals for antibodies to the bacterial agent responsible for the disease, researchers were able to identify regions posing the greatest danger of infection. *Science News*, Vol. 152, No. 5, August 2, 1997, p. 72–73.

  Research the theory behind landscape epidemiology and describe how it could be used to predict outbreaks of infectious diseases, such as Lyme disease, choler, or leishmaniansis.

ON THE JOB

STATISTICIAN There was a constant hum in the office. Bureau workers were gathering incoming data and watching readouts on their computer screens that were scrolling numbers and demographics of millions of people. Louis was excited to start working with the new information and to see how accurate and efficient the latest survey process was.

The United States census, instituted in 1790 by George Washington, was to be carried out once every 10 years. That first year, the information requested consisted of the number of people living in the country. Little was asked about the inhabitants except their ages and who the head of the household was. The information was to be used primarily for two purposes: to allocate seats in the

House of Representatives (the number of seats depending on the population in a state) and to provide a record of the country's military potential (at that time, males over the age of 16) in case of war.

Much has changed since the beginnings of the process, from technology to (most obviously) the number of people in the United States. Two hundred years after the first census, the information requested of the growing, diversifying population had increased to include wealth, race, education, means of transportation, etc. When all of the information was collected from the 1990 census- takers, it was time for Louis to begin using the data to create some understanding of the U.S. population. How many people lived in the U.S. and its territories? What was the poverty rate? How many children were growing up in single-parent households? What state had had the greatest influx of residents since 1980?

Because the collection of data is so tedious and the U.S. population is so mobile, Louis and the other statisticians with the Department of Commerce must use the returned data (sometimes coming from only 50% of an area's inhabitants) and, with different statistical techniques and formulas, create a profile of the U.S. population that is as close as possible to reality. This information is in demand from many different public and private enterprises, not only for its present value, but for forecasting future trends. For example, if it looks as though more and more people are moving into particular areas, how can those areas accommodate the increased traffic, demand for resources, etc?

Louis has always enjoyed math. He took part in his high school's math club, attending many contests where he was thrilled to compete with his peers. It was only natural for him to follow a mathematical path while attending college. He majored in statistics and rounded out his education with a minor in sociology, reflecting his interest in people. While attending a job fair before graduation, he spoke with a representative from the Census Bureau and was excited by the possibility of a career that included both of his interests.

Between the actual census years, Louis spends his time working with the latest data, analyzing and arranging it for different uses, and studying ways to make the next census even more accurate. Not only has Louis enjoyed his job because of the mathematics involved, but also because of his acquired knowledge about the U.S. population, an ever-changing, evolving, mobile society.

VARIATION...ADAPTATION ...EVOLUTION

PROLOGUE **S**o far in this module, you have been exploring the interactions among organisms and the relationships of organisms to their environments. Living ecosystems are remarkably complex with a variety of plants and animals of different sizes and shapes living distinct but connected lives. Why is there such an assortment of organisms? How did this diversity come about? What might we expect in the future?

Between the years 1849 and 1860, the English naturalist Henry W. Bates wandered through the forests of the Amazon capturing and identifying butterflies. After grouping the butterflies having similar characteristics together, Bates found he had 94 separate types. We now know that there are more than 100,000 different types of butterflies and moths on Earth, which makes them the second largest group of insects. (The largest group is that of the beetles, with over 250,000 different types.) Why are there so many different kinds of insects? Does each kind have a separate niche in the ecosystem that is Earth?

How could such a great diversity of life on Earth come into existence? In this learning experience, you will explore the principles of *evolution* (changes in populations of organisms over time) and, in particular, Charles Darwin's *theory of natural selection.* Over the century and a half since Darwin published *The Origin of Species*, evolutionary biologists have carried out intense research into how organisms evolve. Although they may not be able to explain all the processes that allow evolution to occur, the general process of evolutionary change is accepted by scientists, just as gravity is accepted even if physicists can not completely explain it either.

WHAT IS THIS, ANYWAY?

INTRODUCTION In recent years, new products have been introduced, such as compact discs and cellular phones. In many basements or garages, people are "tinkering," trying to invent a new device that can be patented and produced and will, hopefully, make them rich. An important consideration for inventors is that the structure (shape, size, features) of the object must be related to the function it will perform. For example, if you invent a high-tech, futuristic lamp, it still needs to carry out its original function of delivering light. Similarly for organisms, the structures must help them carry out their life processes, from food-getting to reproduction, and help them function in their unique environments. In the following activity, you will explore the relationship of structure to function, an integral concept to the continued existence of all organisms.

TASK

1. Number a lined sheet of paper according to your teacher's instructions.
2. Go to your assigned "station" and observe the mystery object. What do you think its function is? Write your response next to the corresponding station number on your sheet.
3. At your teacher's signal, move to the next station. What is the function of this object? Write your response next to the proper number on your sheet.
4. Continue, at the signal, until you have examined all the objects and written your predictions as to their functions.

ANALYSIS

Write responses to the following in your notebook.

1. What general principle or concept can you state after doing this activity?
2. What might be the importance of this principle in the plant and animal world? Explain by using several examples.
3. In the natural world of organisms, which do you think might come "first," the structure or the function? Explain your response.

EVOLUTIONARY THEORY: PAST AND PRESENT

When Charles Darwin (see Figure 5.1) set sail from England in December 1831 as the naturalist aboard the H.M.S. *Beagle*, he was a twenty-two-year old university graduate who didn't yet know what work he wanted to do. By the end of his five-year journey (shown in Figure 5.2), he had collected many specimens from around the world and had made voluminous observations of the flora and fauna in South America. He pondered the mysteries of sea shells found thousands of feet up in the Andes mountains of South America; of 68 different kinds of beetles collected in just one day; and of remote islands where the birds and tortoises were similar, yet different both from others of their kinds and from those on the mainland. Darwin would spend the rest of his life as a biologist—though he could scarcely have known he would change how people viewed the living world.

What could have caused all this diversity of life that Darwin observed? For over twenty years after his return in 1836, Darwin reviewed his observations, read widely, and formulated his explanation about how organisms change over time. He published, in 1859, *The Origin of Species by Means of Natural Selection*, in which he set forth ideas that would revolutionize scientific thinking and influence how we look at life on Earth.

Figure 5.1
Charles Darwin shortly after he returned from his voyage on the H.M.S. *Beagle*.
Photo Researchers, Inc.

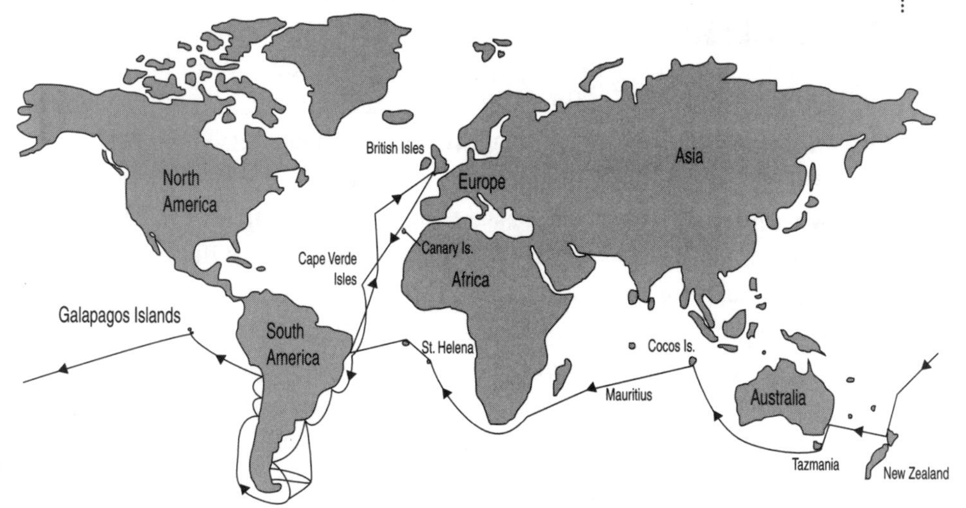

Figure 5.2
Route taken by the H.M.S. *Beagle*.

His *theory of natural selection* includes the following precepts:

- Organisms reproduce others of their own kind.
- In nature, there is an overproduction of offspring.
- There are variations (differences) among offspring and some of these variations will be inherited.
- Organisms having "favoured" variations, such as longer legs (for speed) or outer coloring (for camouflage) are more likely to survive and pass on these *adaptations* to their offspring ("the survival of the fittest").
- Nature "selects" those organism whose adaptations allow them to survive in their environment, thereby enabling them to reproduce.
- Over time, the survivors with the favorable adaptations will make up most of the population.

Charles Darwin's ideas were influenced by those of scientists and others, such as plant and animal breeders who used artificial breeding to produce better crops and livestock. The breeders informed him that there were variations in such crops as corn and in the amount of milk that cows produced. From Charles Lyell's *Principles of Geology*, Darwin learned that the earth was much older than previously thought. From Thomas Malthus' essay on populations, Darwin realized the importance of the idea that more offspring are born than can live. For example, a tree produces thousands of seeds, but only the few that fall where conditions are favorable can survive.

Darwin wondered what other factors might determine which offspring survive and reproduce. Perhaps such survival was due to the variations mentioned by the plant and animal breeders. Although he was writing *The Origin of Species* at the same time as Gregor Mendel was beginning his genetics experiments on pea plants, Darwin did not understand the cause of the variations on which his theory was built. He realized that organisms had characteristics or traits, such as coat or flower color, but he did not understand genes or DNA or how these traits were inherited. After Mendel's paper was rediscovered by later biologists in 1900, some of the reasons for variation among offspring became understood. They were due to the action of dominant and/or recessive "factors" (now known as genes). Subsequent research identified mutations, or changes in the DNA as causes for variation. Continuing research has illuminated the biological basis for variation, a central tenet of natural selection.

Some of Darwin's ideas have been modified since 1859, but his basic ideas that organisms undergo evolutionary change and are selected by their environments remain true and are undeniable. It is important to note that while individuals and their descendants vary, they, themselves, do not evolve; it is the accumulation of many changes in a population (due to environmental pressures) that results in the evolution of one

Figure 5.3
Beaks of Darwin's finches.

species of organisms into another over time. The accumulation of these small changes over millions of years is referred to as *gradualism.* But not all changes among living things are slow and gradual; some are much more abrupt. It is thought that environmental pressures determine the pace of evolution, gradual at times and more rapid at others. Current yearly research by Peter and Rosemary Grant on the Galapagos island of Daphne Major, for example, has shown that the beaks in the finch populations are continually evolving as their food changes in response to climatic variations. This is evolution that can be followed in a human lifetime.

The *fossil record* (shown in Table 5.4)—the cataloging and describing of fossils—is a constructed history of life on Earth, divided into eras, periods, and epochs. It also shows that after long periods of relative sta-

Table 5.4
Geologic time table and major life forms.

ERA	PERIOD	EPOCH	YEARS (Millions)	LIFE FORMS
Cenozoic	Quaternary	Recent	0.01	Modern humans, modern plants and animals
		Pleistocene	2.5	Early humans, extinction of many large mammals and birds
	Tertiary	Pliocene	7	
		Miocene	26	Whales, apes, grazing mammals, spread of grasslands
		Oligocene	38	Monkeylike primates, land dominated by flowering plants
		Eocene	54	
		Paleocene	65	
Mesozoic	Cretaceous		136	Flowering plants, extinction of dinosaurs
	Jurassic		190	Dinosaurs dominant, primitive birds appear, flying reptiles and small mammals, conifers
	Triassic		225	Dinosaurs and early mammals, primitive seed plants
Paleozoic	Permian		280	Rise of insects, reptiles diversify
	Carboniferous		345	Insects and amphibians, mosses and ferns; first reptiles, variety of insects
	Devonian		395	Age of fish; early amphibians; mosses, liverworts, and ferns
	Silurian		430	First land plants, early arthropods, rise of fishes and reef building corals
	Ordovician		500	Primitive mollusks and fish, marine algae
	Cambrian		570	Sponges, jellyfish, worms, primitive algae
Precambrian			4500	Monerans, simple protists

bility, during which very little change occurs over a species' lifetime, there are periods of abrupt and dramatic change lasting perhaps tens of thousands of years (which is "sudden" in geologic time). During these periods, current species change and/or die out, and new species emerge. This pattern of periods of almost no change followed by a rapid burst of evolution is called *punctuated equilibrium*, or periods of slow change, interrupted by periods in which rapid change occurs.

The complete disappearance of one type of organism (or *extinction*) results from extreme environmental pressures, human or natural (see Figure 5.5). The causes of most of these extinctions are unknown, but there is scientific evidence that an enormous asteroid crashing on the Yucatan peninsula of Mexico about 65 million years ago caused the extinction not just of the dinosaurs, but of nine-tenths of all species then

Figure 5.5
Mass extinctions.

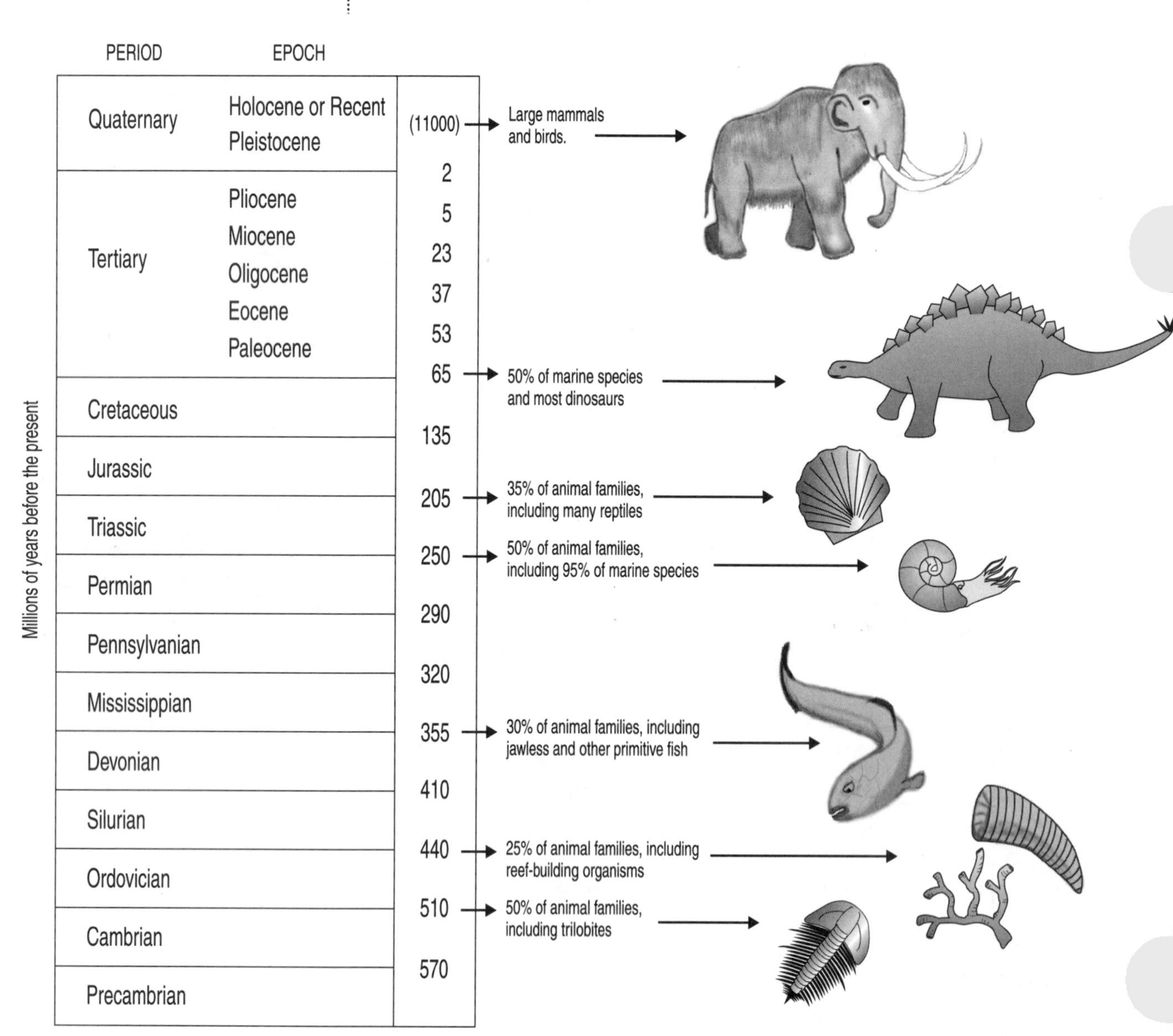

living on the earth. The asteroid caused a huge dust cloud to form, blocking sunlight and resulting in climatic changes—specifically, the cooling of Earth. Thus, the dominant plants died out and so did the animals that depended on them for food, including the dinosaurs.

Organisms that have disappeared over time were not "inferior" in any sense to those that survived. They were just less lucky. The history of life has been profoundly influenced by long-term evolutionary interactions where key changes in structure and life cycle provide those variations that are more likely to survive in a changing environment. Although natural selection is a potent force some species become extinct by chance or by human intervention, not because they were poorly adapted

ANALYSIS

Write responses to the following in your notebook.

1. Explain, in your own words, Darwin's theory of natural selection. Relate it to the concept of evolution or change over time.
2. Look at the beaks of the finches that live on the Galapagos Islands (Figure 5.3). What type of food do you think each finch eats? Why do you think so?
3. Explain the role of the environment in both evolution and in extinction.
4. Compare gradualism and punctuated equilibrium. Give an example of each.
5. How might punctuated equilibrium explain gaps in the fossil record?

GOING WITH THE FLOW

INTRODUCTION In order for evolution to occur there must be variations among individuals, for it is these variations that allow populations to change as the environment changes. Do these variations come about in order to ensure survival, or are they the unplanned consequences of changes in the DNA and of random mating? In the following simulation, you will explore changes in individuals that make up a population.

MATERIALS NEEDED

For each pair of students:

- 1 paper bag with movement directions
- 2 sheets of graph paper
- 2 pens or pencils (of different colors)

▶ PROCEDURE

1. Fill in the center square of the graph paper grid with a pencil.
2. Shake the paper bag to mix the contents.
3. Without looking in the bag, draw out a single strip of paper. Fill in your grid according to the instruction on the strip of paper (see Figure 5.6).
4. Place the strip of paper back in the bag. Repeat this sequence 20 times, using the same pencil.
5. Circle the last square you filled in and wait for your teacher's instructions.
6. If your "organism" did not survive, join a pair whose organism did survive.
7. Continue, beginning with the last (circled) square, and follow procedure steps 2-5 using a different color of pencil or pen.
8. Observe the grids of the other groups.

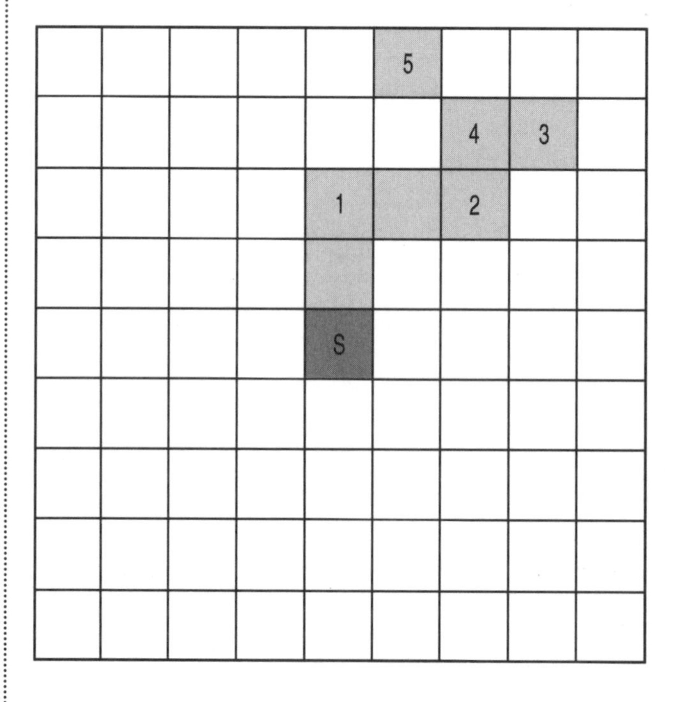

Figure 5.6
Sample random moves.

▶ ANALYSIS

Write responses to the following in your notebook.

1. What does the central square of the grid signify?
2. In evolutionary terms, what were you showing as you followed the directions on the strips?
3. Why did some grids not survive?

4. Explain why your completed grid was different from the others. Use your understanding of Darwin's theory of natural selection.
5. How might this illustrate the principle that individuals vary but do not evolve, that populations evolve?

Of Fish and Dogs

INTRODUCTION Throughout this module you have investigated different populations of organisms: those in your square meter, others in food webs, and still others in your study of growth curves and biogeochemical cycles. As you read in the Prologue of this learning experience, Henry W. Bates grouped butterflies of similar appearance together and called them separate species, as did most collectors of the time. But is there more to the concept of species than meets the eye?

Consider the cichlids, perch-like fish found in the fresh waters of South America, Africa, Sri Lanka, and India. Cichlids have been of great interest to researchers for several reasons; one of them is their parenting behavior. Cichlids hatch their young in their mouths (mouth brooding), and both parents show strong nurturing and protective ties to the young, behavior unusual for fish.

Dogs are… well, dogs.

In the following activity you will examine diagrams that show a collection of cichlids found in Lake Victoria in Africa, and a collection of dogs. How many different species can you identify in each illustration?

TASK

Examine the two groups of animals shown in Figure 5.7 and Figure 5.8 on the following pages, and determine how many different species you think are in each group. List the reasons (criteria) that you used to separate species within each collection. Be prepared to explain them to the rest of the class.

Figure 5.7
Dogs: how many different species?

Figure 5.8
Cichlids: how many different species?

READING

The Fundamental Unit

Reprinted by permission of the publisher from THE DIVERSITY OF LIFE *by Edward O. Wilson, Cambridge, Mass.: Harvard University Press, Copyright © 1992 by Edward O. Wilson, pp. 38–39.*

...I will try to cut to the heart of the matter with the "biological-species concept": a species is a population whose members are able to interbreed freely under natural conditions. This definition is an idea easily stated but filled with exceptions and difficulties, all interesting...I must add at once that not all biologists accept the biological-species concept as sound or as the pivotal unit on which the description of biological diversity can be based. They look to the gene or the ecosystem to play these roles ...I think they are wrong, but in any case will return shortly to the difficulties of the biological species to give voice to their misgivings.

For the moment let me go on to expand the definition, which is accepted at least provisionally by a majority of evolutionary biologists. Notice the qualification it contains, "under natural conditions." This says that hybrids [that is, the descendants of different kinds of animals] bred from two kinds of animals in captivity, or two kinds of plants cultivated in a garden, are not enough to classify them as a member of a single species. To take the most celebrated example, zookeepers have for

Continued on next page

years crossed tigers with lions. The offspring are called tiglons when the father is a tiger and ligers when the father is a lion. But the existence of these creatures proves nothing, except perhaps that lions and tigers are genetically closer to each other than they are to other kinds of big cats. The still unanswered question is, do lions and tigers hybridize freely where they meet under natural conditions?

Today the two species do not meet in the wild, having been driven back by the expansion of human populations into different corners of the Old World. Lions occur in Africa south of the Sahara and in one small population in the Gir forest of northwestern India. Tigers live in small, mostly endangered populations from Sumatra north through India to southeastern Siberia. In India, no tigers are found near the Gir forest. It would seem at first that the test of the biological-species concept, free interbreeding in nature, cannot be applied. But this is not so: during historical times the two big cats overlapped across a large part of the Middle East and India. To learn what happened in these earlier days is to find the answer.

At the height of the Roman Empire, when North Africa was covered by fertile savannas—and it was possible to travel from Carthage to Alexandria in the shade of trees—expeditions of soldiers armed with net and spear captured lions for display in zoos and in coliseum spectacles. A few centuries earlier, lions were still abundant in southeastern Europe and the Middle East. They preyed on humans in the forests of Attica while being hunted themselves for sport by Assyrian kings. From these outliers they ranged eastward to India, where they still thrived during British rule in the nineteenth century. Tigers ranged in turn from northern Iran eastward across India, thence north to Korea and Siberia and south to Bali. To the best of our knowledge, no tiglons or ligers were recorded from the zone of overlap. This absence is especially notable in the case of India, where under the British Raj trophies were hunted and records of game animals kept for more than a century.

We have a good idea why the two species of big cats, despite their historical proximity, failed to hybridize in nature. First, they liked different habitats. Lions stayed mostly in open savanna and grassland and tigers in forests, although the segregation was far from perfect. Second, their behavior was and is radically different in ways that count for the choice of mates. Lions are the only social cats. They live in prides, whose enduring centers are closely bonded females and their young. Upon maturing, males leave their birth pride and join other groups, often as pairs of brothers. The adult males and females hunt together, with the females taking the lead role. Tigers, like all other cat species except lions, are solitary. The males produce a different urinary scent from that of lions to mark their territories and approach one another and the females only briefly during the breeding season. In short, there appears to have been little opportunity for adults of the two species to meet and bond long enough to produce offspring.

Thus, Wilson's criterion of a *species* is whether members of that group can interbreed. This definition has been extended by others to include the idea that this interbreeding must result in the production of fertile offspring, that is, offspring capable of bearing offspring themselves. In some instances, this is physically impossible. Size and structure are not the only physical limitations; even organisms of similar structure may have differing numbers of chromosomes which would make production of a viable embryo impossible.

Even in cases in which it is physically possible for hybridization between two species to occur, it will not usually occur in the wild. As you have seen from Wilson's description of tiglons and ligers, the definition of species is complex and the key term is "under natural conditions." Habitat isolation is another reason two species capable of breeding may not do so; if two organisms are not introduced, they have no chance to court and mate. And even if they do meet, other features such as species-specific coloration, scent, variation in reproductive timing (such as when the female is in heat), and courtship behavior will indicate to potential suitors that the mate in question is of a different species and not an appropriate mate. Even though all cichlids in Lake Victoria may look pretty much alike to us, the fish themselves are clearly quite aware of which fish is which, and will only breed with fish of the same species.

ANALYSIS

Write responses to the following in your notebook.

1. Explain why cichlids are considered different species, but dogs, the same species. Be sure to include in your explanation factors which would or would not influence hybridization.
2. How do you think the concept of separate finch species is related to separate cichlid species?
3. The term "hybrid" is often used to describe the offspring of different species which have been bred under artificial circumstances in laboratories, zoos, or farms. A technique called crossbreeding allows the breeder to produce new organisms with very specific traits. An example of such crossbreeding is the mating of a horse and a donkey which produces a mule. A horse (*Equinus caballus*), which has 64 chromosomes, will mate with a donkey (*Equinus asinus*), which has 62 chromosomes; their mule offspring will have 63 chromosomes and will be sterile. Explain what you think this means in terms of species. How does this extend your understanding of the definition of species?

CICHLIDS PAST AND PRESENT

The cichlid fish of Lake Victoria in East Africa are remarkable not only for their variations in color and in their parental attentiveness as mouthbrooders, but also for their ancestral origins. Within this one (albeit very large) lake live 300 hundred or more distinct species of cichlids. While distinctive in color, structure and feeding habits—some are bottom-feeding algae-eaters, some eat snails, others mollusks, others fish, some only eat fish scales, and still others prefer only the eyes of fish—the cichlids are astonishingly the same at the level of their DNA, differing in some cases by only one nucleotide. All of the different species of cichlids in Lake Victoria have evolved from a single ancestral species within the last 750,000 years. In 1959, British colonists introduced the Nile perch, a sport fish which grows to almost two meters in length, to the lake. A voracious predator, this fish has single-finedly (sic) undid thousands of years of evolution in a single gulp by decimating the cichlid population and eliminating many species. In doing so, this large fish has also upset the ecosystem balance of the lake by eliminating the algae-eating cichlids. In the absence of the cichlids, algal blooms fill the lake. When these blooms die and decompose, the oxygen content of the lake is reduced, which results in a decline in the population of other fish, crustaceans and other biotic members of the ecosystem.

EXTENDING IDEAS

- If you are fascinated by the extinction of the dinosaurs, you may wish to read *T. Rex and the Crater of Doom* by Walter Alvarez (Princeton, NJ: Princeton University Press, 1997). This is a highly readable scientific book that gives insights into how discovery leads to more riddles that need to be solved and shows the dynamic nature of research and of the personalities of the scientists who are involved in the mystery of the death from above.

ON THE JOB

ORNITHOLOGIST Leah cringed on her couch, hiding her head behind a pillow until the credits to Alfred Hitchcock's "The Birds" scrolled up the TV screen. The fictitious birds in the movie were stalking the residents of a small town, waiting on rooftops, power lines, and other perches before attacking them. Although Leah was terrified throughout the film, she found herself greatly interested in the birds—so interested that this film led Leah into a career as an ornithologist.

When Leah saw "The Birds," she was in high school. She was sure that birds did not act as they did in the film, but she wanted to learn more about them. After visiting the library and thumbing through numerous books on the subject, she decided to change her path in school so that she could continue to learn more about birds and their behavior. That meant adding more science courses, especially biology, to her course load. She joined the National Audubon Society and sometimes found time on weekends and after school to volunteer at a local bird sanctuary. There she clerked in the bookstore, filled the bird feeders, and took notice of all of the different species of birds that visited the feeders as well as their unique behaviors.

Leah's family was very proud when she was accepted into a top university and when, four years later, she graduated with a Bachelor of Science degree in wildlife management. Not wanting to stop at an undergraduate degree, she continued on to get a Master's Degree in environmental biology.

Having already been involved with the National Audubon Society's Christmas bird counts, Leah became a bird researcher, spending most of her time on a study of the piping plover. She and her coworkers had the goal of increasing the population of this shore-nesting bird by protecting their nesting sites from human encroachment. Many plover nests, in the past, have been tampered with or destroyed because of automotive and all-wheel drive beach access. Leah's days were filled by trips to the beaches to check the status of the nesting pairs. Sometimes she was able to visit other nesting sites on the East Coast to compare notes with other researchers. During these visits, she would count and measure eggs, make sure that nesting sites were cordoned off from the public (sometimes that meant talking with town or city personnel about closing off an entire beach, which often dismayed summer residents and tourists), and counting the nesting pairs. This might mean banding birds that hadn't yet been identified. There was a marked increase in the numbers of nesting pairs within a few years after she started with the project. During the winter months after the piping plovers moved south, Leah spent a lot of time in front of a computer, analyzing her data and getting ready for the next nesting season.

Even though her job kept her very busy, Leah still made time to volunteer at her local bird sanctuary, still filling bird feeders, but also leading seminars to educate the public about their feathered friends and how best to live harmoniously among them.

THE DIVERSITY OF LIFE

LEARNING EXPERIENCE 6

PROLOGUE You have been introduced to the diversity of life on Earth throughout this module: in your field study, in food web explorations, in your ecocolumn, in population studies, and in your readings. There are many aspects of diversity, but in most cases the term refers to the number of species on Earth.

Scientists and naturalists have described about 1.8 million species of living organisms and estimate that there are 5 to 100 million in all. Most of the undiscovered kinds of life are thought to be in the ocean depths, in the soil, and in tropical rain forest canopy. Sometimes, species are "discovered" by scientists, but they are already known to the native population (an example is the tree kangaroo of New Guinea "found" in 1994). Figure 6.1 shows a representation of the number of species among life forms. Insects, for example, account for half of all known species (approximately 800,000 species). The larger forms of life have been described and classified more completely, although they make up only a small percentage of the total number of species. Why is the picture of the elephant so small, and that of the insect so large?

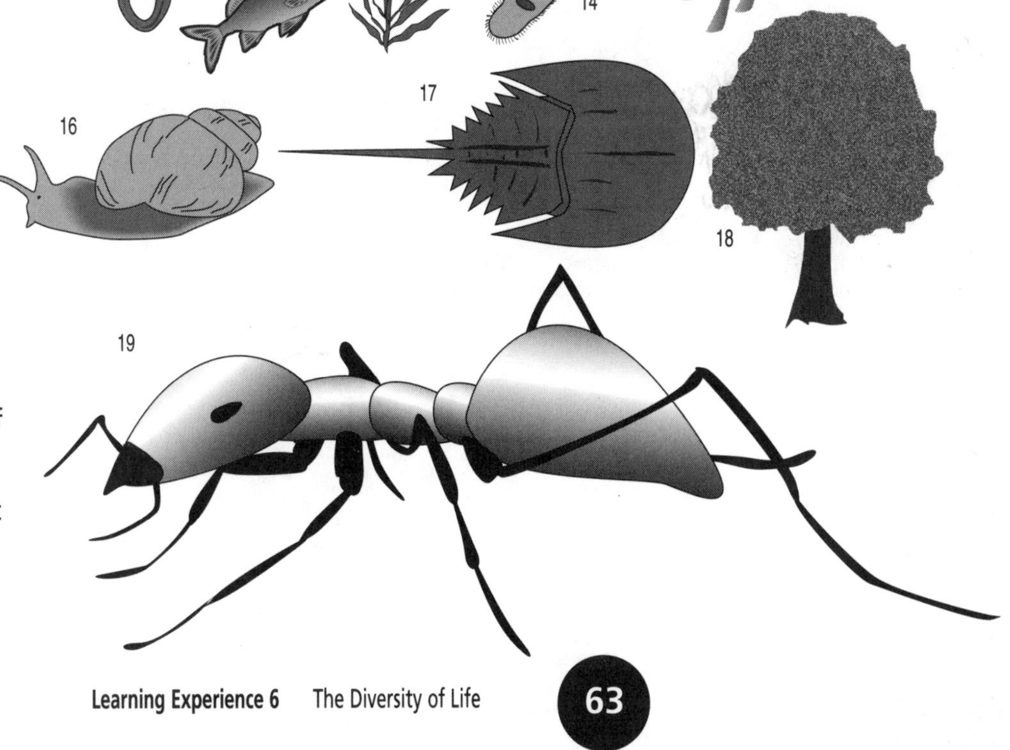

Figure 6.1
Each organism represents a major group, of species. For example, the ant (19) represents all the insects in the world.

As you learned in the last learning experience, many more millions of species that once lived on earth have died out. But how have new species appeared (*speciation*)? What can explain so many different species of cichlids or so many changes in species over time? In this learning experience, the answers to these questions will become clearer as you explore the origin of and the changes within the diversity of life.

READING

New Species

Reprinted by permission of the publisher from THE DIVERSITY OF LIFE *by Edward O. Wilson, Cambridge, Mass.: Harvard University Press, Copyright © 1992 by Edward O. Wilson, pp. 51–74.*

What is the origin of biological diversity? This profoundly important problem can be most quickly solved by recognizing that evolution creates two patterns across time and space. Think of a butterfly species with blue wings as it evolves into another species with purple wings. Evolution has occurred but leaves only one kind of butterfly. Now think of another butterfly species, also with blue wings. In the course of its evolution it splits into three species, bearing purple, red, and yellow wings respectively. The two patterns of evolution are vertical change in the original population and speciation, which is vertical change plus the splitting of the original population into multiple races or species. The first blue butterfly experienced pure vertical change without speciation. The second blue butterfly experienced pure vertical change plus speciation. . . The origin of most biological diversity, in a phrase, is a side product of evolution.

Vertical change is mostly what Darwin had in mind when he published his 1859 masterwork. The full title tells the story: *On the Origin of Species by Means of Natural Selection, or the Preservation of Favoured Races in the Struggle for Life.* In essence, Darwin said that certain hereditary types within a species (the "favoured races") survive at

Figure 6.2
Vertical evolution. A blue-winged butterfly species may evolve over time and space and become purple-winged.

the expense of others and in so doing transform the makeup of the entire species across generations. A species can be altered so extensively by natural selection as to be changed into a different species, said Darwin. Yet no matter how much time elapses, no matter how much change occurs, only one species remains. In order to create diversity beyond mere variation among the competing organisms, the species must split into two or more species during the course of vertical evolution.

… Any evolutionary change whatsoever that reduces the chances of producing a fertile hybrid can yield a new species. . . Consider a male of species A and a female of species B trying to create a fertile hybrid offspring. Because they are genetically different from one another, things can go wrong. The two individuals might want to mate in different places. They might try to breed at different seasons or times of the day. Their courtship signals could be mutually incomprehensible. And even if the representatives of the two species actually mate, their offspring might fail to reach maturity, or attaining maturity, turn out to be sterile. The wonder is not that hybridization fails but that it ever works. *The origin of species is therefore simply the evolution of some difference—any difference at all—that prevents the production of fertile hybrids between populations under natural conditions.*

… But wait: I have been speaking of the origin of species in paradoxical language. In the traditional language of biology, the "mechanisms" have "functions." Yet they represent whatever can go wrong, not what can go right. In other words, beauty arises from error. How can both of these apparently contradictory perceptions be true? The answer, based on studies of many populations in the wild, is this: *the differences between species ordinarily originate as traits that adapt them to the environment, not as devices for reproductive isolation.* The adaptations may also serve as intrinsic isolating mechanisms, but the result is accidental. Speciation is a by-product of vertical evolution.

To see why this strange relationship holds, consider the special but widespread mode of diversification called *geographical speciation.* Start with an imaginary population of birds—say, flycatchers—that was split by the last glacial advance in North America. Over several thousand years, the population living in what would today be the southwestern United States adapted to life in an open woodland, while the other population, in the southeastern United States, adapted to life in swamp forests. These differences were independently acquired and functional. They allowed the birds to survive and reproduce better in the habitats most readily available to them south of the glacial front. With the retreat of the ice, the two populations expanded their ranges until they met and intermingled across the northern states. One now breeds in open woodland, the other in swamps. The differences in their preferred habitats, based on hereditary differences acquired during the period of enforced geographical separation, makes it less likely that the two newly evolved populations will closely associate during the breeding season and hybridize. The adaptive difference in habitat thus accidentally came to serve as an isolating mechanism.

… The two populations have turned into distinct species because they are reproductively isolated where they meet under natural conditions. The single ancestral, pre-glaciation species has been split into two species, an entirely incidental result of the vertical evolution of its populations while they were separated by a geographical barrier.

Figure 6.3
Divergent Speciation. A blue-winged butterfly species may evolve over time and space into purple-, yellow-, and red-winged species.

… Great biological diversity takes long stretches of geological time and the accumulation of large reservoirs of unique genes. The richest ecosystems build slowly, over millions of years. It is further true that by chance alone only a few new species are poised to move into novel adaptive zones, to create something spectacular and stretch the limits of diversity. A panda or a sequoia represents a magnitude of evolution that comes along only rarely. It takes a stroke of luck and a long period of probing, experimentation, and failure. Such a creation is part of deep history, and the planet does not have the means nor we the time to see it repeated.

► ANALYSIS

Write responses to the following in your notebook.

1. How would you define speciation?
2. Create a labeled flow chart that diagrams and explains Wilson's two examples about the blue butterflies.

3. How do geographic isolation and reproductive isolation lead to speciation? Use Darwin's finches as an example in your response.
4. Do you think speciation is evolution? Explain your response.
5. Discuss the meaning of the quotation "The environment is the theatre and evolution is the play."

Mapping the Gradient Across the Americas

INTRODUCTION The term *biodiversity* (a contraction of biological diversity) in part refers to the variety of life forms on Earth. Biodiversity, often referred to as *species richness*, is the number of species in an area or habitat. Confronted with the vast numbers and diversity of species, one might ask several questions: What is the distribution of various species? What are the factors that influence where plants and animals can be found? How are evolution and speciation related to biodiversity?

In this activity, you will identify patterns in the distribution of species and determine general principles that help explain species distribution.

MATERIALS NEEDED

For each group of three students:

- 1 set of "Species Richness Maps"
- assorted colored pencils

For the class:

- reference material including atlases, encyclopedias, and geography books

PROCEDURE

1. Color each map so that patterns in each species become more apparent. For example, leave all the spaces between lines 0–99 white, color red between lines 100–199, etc.
2. Create a color code, place it in the corner of the sheet for reference, and use the same code in each map.
3. Discuss the Analysis with your group members.

▶ ANALYSIS

Write responses to the following in your notebook.

1. List the areas in each map having the greatest number of species. List those with the fewest.
2. Analyze your lists. Which of the areas are the same? Which are different? What might be some reasons for this?
3. Examine topographical, land form, and climatic regions maps of North and Central America. What patterns do you see that might explain the reasons for greater or lesser species diversity?
4. List several general principles that explain how geography plays an important role in species diversity.
5. Examine Figure 6.4. Using concepts from this module, write a short essay which explains the reasons for and possible consequences of the changes shown in Costa Rica.

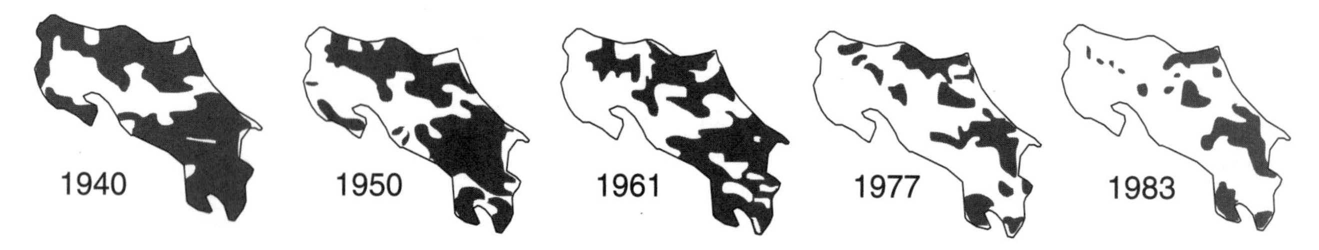

Figure 6.4
Forest areas (dark) in Costa Rica from 1940 to 1983.

EXTENDING IDEAS

- Investigate and visit your local branch of the National Audubon Society and/or other nature/conservation societies in your area. They sponsor many events including bird watching, nature walks, and other outdoor activities.
- In *After Man: Zoology of the Future* (New York: St. Martin's Press, 1981), Dougal Dixon has created an interesting "zoo of the future." He contemplates future evolution on our own planet. He waves a time wand, sets his scenario 50 million years from now, eliminates today's dominant species, and creates new animals that might take over as the major occupants of the earth's surface. He uses them to show some of the basic principles of evolution and ecology. He does not make firm predictions that this is what the animals would look like, but offers an exploration of the possibilities.

  Design your own "critter" (herbivore or carnivore) that might appear in the future. Place it in its specific environment, describe its niche, and explain how it is adapted to its environment.

- Species and groups of related species are emerging while other species are dying out. Evolution is a dynamic process. On Madagascar, an island in the Indian Ocean off the east coast of Africa, live 30 species of lemurs (a type of primate found nowhere else in the world). These 30 originated from one or two ancestral species. Research the evolution of lemurs and the speciation that resulted from changes over time.
- The history of humans on earth is an example of evolution, speciation, and extinction. The earliest hominid known from the fossil records is *Australopithecus afarensis*, which lived in Africa about three to five million years ago. This species, thought to be the only hominid species around at that time, was 1.4 meters tall, walked upright (was bipedal) and possibly lived in family groups. Their cranial capacity (brain size) was about the same as the modern day chimpanzee, and their teeth remains indicate they were mostly vegetarian. About two million years later speciation of hominids occurred, resulting in at least three distinct species. Research and diagram human evolution from *Australopithecus* to modern day *Homo sapiens sapiens*.

ON THE JOB

CARTOGRAPHER "Marsha's Maps. May I help you?"

"I hope so. I am interested in adding a detailed, interactive road map of Hopedale, showing street names and major building landmarks, to our town's web page."

"Well, you've contacted the right place!"

Marsha is a cartographer. A few years ago, she started her own business. With recent leaps in technology, she has been able to cater to a larger clientele than she had ever dreamed, and she has created maps for magazines, tourism guides, government agencies, municipalities, and World Wide Web sites.

Marsha's interest in maps began with the treasure maps that her father created for her when she was a child. She soon began manually mapping out her neighborhood, including all of the popular hangouts and spots of interest (such as the local burger joint).

In high school, Marsha took courses in algebra, geometry, trigonometry, drafting, mechanical drawing, and computer science to help her toward her goal of becoming a cartographer. Once out of high school, she began working toward a four-year degree in engineering at the university. When her financial aid fell through after two years, she became concerned that she would not reach her goal, but Marsha learned that she could still get an associate's degree from a two-year college, then become an apprentice to an established mapmaker, and still become a cartographer. With technological advances, she also knew that her computer training would be an important aspect of her career.

She began her career as a surveyor, doing hands-on data collecting using theodolites (tools used to measure horizontal and vertical angles) and electronic equipment used to measure distances. Sometimes she even found herself using a simple tape measure.

After working for a number of years and getting an adequate feel for the field of cartography, Marsha felt ready to take the data and create her own maps in her own company. Her maps have included relief maps, which show land forms such as hills, mountains, and trees from above; city maps, which she finds challenging because of the amount of information needed to fit clearly into a small space; tourism maps, which need to be informative, yet make an area inviting to visitors; and many others for special projects. Her data usually come from digital data received from Geographic Information Systems (GIS) and, for much larger projects, from the Global Positioning System (GPS), is a satellite system that can precisely locate points on the earth using radio signals transmitted for satellites. These systems have made it possible for Marsha to create interesting, informative, and artistic maps using little more than her computer and her knowledge of surveying and mapping the earth. She works closely with clients, making sure to create exactly what they had expected before hiring her.

One recent client asked her to create a map of some hiking trails in a popular state park. Because this information was not readily available, she hired a consultant surveying company to help her. By carrying handheld systems, the surveyors were able to receive GPS signals as they walked and rode over the trails, locating each point along the paths and storing the information for Marsha to use. Combining the trail data with existing maps of the state park area, Marsha was able to create a layered map of the trails, hills, and overhanging trees. This type of information not only made it possible to create printed maps of the trails for visitors, but also could be extremely helpful to the park rangers should they ever need to locate someone who was lost on the trails.

"Marsha," called one of her associates from her work station. "Hopedale's web site is active and your map looks spectacular! The mayor called and said he has already received wonderful feedback from residents and visitors!"

Back to Nature

PROLOGUE Is it possible to reverse the current trend of destroying natural habitats and the organisms in them? For many years people have tried to save the environment by conserving resources and reducing pollution, but until recently, few people were thinking about recovering lost ecosystems. Environmentalists, politicians, and businessmen are now working on restoring selected environments. The goal in *restoration ecology* is for the habitat and its resources to be physically repaired, and, if necessary, its missing components replaced. There are many types of restoration projects. Some projects improve depleted farmland soil; some, like the return of the wolves to Yellowstone National Park, restore populations of plant or animal species; others employ techniques for using natural resources without depleting them; and still others convert vacant lots in urban areas into miniparks or gardens.

However, restorations are never perfect reproductions of past ecosystems. The restorations are different because of what is not there any longer—species that have become extinct or have moved to other areas—and also because of what is there now—a bird, insect, or plant newcomer that has invaded the ecosystem and succeeded in making itself at home. As you have learned in this module, even ecosystems that exist "undisturbed" by humans change too, sometimes dramatically. Planners of any restoration project must, therefore, choose both the place and the time in the past that they are attempting to restore.

A major project is now underway in Florida to restore the Everglades, but the outcome is still uncertain. Upon completion, this project will be the largest water-system restoration in history. In this learning experience, you will explore the concept of restoration ecology—and what this could mean for the future—by looking into the issues that surround the Everglades Restoration Project. The story of the Everglades reveals the complex relationships between humans and the environment.

Anything We Want

INTRODUCTION Humans have disturbed or damaged of Earth's ecosystems. Yet, with our current ecological knowledge and experience it may be possible to restore or repair some ecosystems. Restoring an ecosystem that has deteriorated is complicated. First you need to know what the original looked like and how it functioned. Then you need to determine what might be missing, then you need to plan, and finally, to pay for the project.

Begin by reading the Task and the article "Bringing Back the Everglades." Research the biological and social issues surrounding the restoration of the Everglades system and reflect on the possibilities of restoring an ecosystem.

▶ TASK

1. With a partner, identify and record a list of all the issues mentioned in the article. These issues are related to ecological principles and to political, social, or economic topics. (It might be useful to create a separate list of groups, individuals, and organizations that are involved in this restoration project.)
2. Research one topic from the list you made, and be prepared to present your findings in a mock "congressional committee hearing" that has been set up to evaluate the status of the project. You and your partner should focus your research on:
   - the ecological principles that need to be taken into consideration including the organisms affected by the project
   - the views of the companies, groups, or individuals affected by your particular issue
   - your views and your analysis of the pros and cons of the issue
   - your own educated opinion of what should be done
3. Develop an opening statement for the committee. The intent of this speech is to share your knowledge and concerns about the issue you and your partner have researched with the committee members. The speech should last about three minutes. As you develop the speech, consider the following:
   - What messages are you are eager to convey to the members of this committee?
   - What other issues affect your issue, and what do you need to know about them?
4. Take careful notes on the issues that other students have researched. Their expertise is needed in order for you to make an informed decision.

Following the committee hearing, there will be a discussion of the issues that were presented and of the status of the Everglades Restoration Project.

5. Finally you will write an independent comprehensive paper in which you describe what you would do if you were on the committee. Include the rationale behind the restoration project, all major issues, the arguments presented, and your own educated opinion. This paper will be evaluated by your teacher.

Bringing Back the Everglades

AMID GREAT SCIENTIFIC AND POLITICAL UNCERTAINTY, ECOSYSTEM MANAGERS IN FLORIDA ARE PUSHING AHEAD WITH THE BOLDEST—AND MOST EXPENSIVE—RESTORATION PLAN IN HISTORY

by Elizabeth Culotta

Reprinted with permission from Science, *June 23,1995: 268, 1688-1690. Copyright 1995, American Association for the Advancement of Science.*

When steamships plied central Florida's Kissimmee River early in this century, passengers on ships traveling [toward each other] would spot each other across the marshes in the morning, then traverse the serpentine waterway for a full day before meeting. But in the 1960s, the U.S. Army Corps of Engineers straightened out the Kissimmee. In the name of efficiency and flood control, they dug 56 miles of straight canal to replace 103 miles of meanders—and destroyed at least 1.2 million square meters of wetlands in the process. The river was once home to flocks of white ibis; today it boasts the cattle egret, accompanying herds of cows grazing on the canal's linear banks.

But at one spot on the central Kissimmee, boats must again follow the twists and turns of the old river channel. The Corps is slowly putting the kinks back into the Kissimmee. By working with the state of Florida to restore the wetlands, they hope to bring back the invertebrates, fish, and, eventually, the wading birds that once nested here. With an estimated price tag of $370 million, this is the most ambitious river restoration in U.S. history.

It is, however, a mere drop in the watershed compared to plans for the rest of south Florida. Over the next 15 to 20 years, at a cost of roughly $2 billion, the Corps and state and other federal agencies plan to replumb the entire Florida Everglades ecosystem, including 14,000 square kilometers of wetlands and engineered waterways. It's an urgent task, planners say. For after decades of drainage, altered water flow, and pollution, the Everglades is dying, and as they go, so goes the region.

If wetlands that once replenished underground aquifers stay dry, cities may face future water shortages. Anoxic conditions threaten fish in Florida Bay, saltwater intrudes into marshes and drinking wells, and wildlife—including

Continued on next page

55 endangered or threatened species—is at risk. "This is not rescuing an ecosystem at the last minute. This is restoring something that has gone over the edge," says George Frampton, assistant secretary of the Department of the Interior and chair of the federal interagency South Florida Ecosystem Restoration Task Force.

More than the ecology of southern Florida is at stake. Wetlands managers from Australia to Brazil are keeping a close eye on the project as they search for ways to restore their own ravaged regions. If planners can pull it off, the Everglades restoration will become a world model, says wetlands expert Joy Zedler of San Diego State University, who notes that most restorations "are the size of a postage stamp compared to the Everglades." James Webb, Florida regional director of the Wilderness Society and a member of the Governor's Commission for a Sustainable South Florida, puts it another way: "If we can't do it in the Ever-glades, we can't do it anywhere."

The overall goal of the restoration is to take engineered swampland riddled with canals and levees and transform it into natural wetlands that flood and drain in rhythm with rainfall. Planners hope the entire ecosystem—plants and animals—will blossom as a result. "Wet it and they will come" is the unofficial motto. But because no one understands all the complex ecology involved, planners must accept a hefty dose of scientific uncertainty. "We really don't know what we're going to get out there," says biologist John Ogden of Everglades National Park. And the Corps and the South Florida Water Management District (SFWMD), cosponsor of the restoration, still haven't come up with a final blueprint for the replumbing.

The other big unknown in the Everglades is political. Would-be rescuers represent a surprisingly broad coalition of interests and money, from federal and state agencies to environmentalists and urban developers, who want a steady water supply. But holding such a diverse coalition together over the planned life of the project will be tricky. Moreover, the steep price tag—of which one third is supposed to come from the federal budget—and extensive federal involvement run counter to Washington's current budget-cutting mood. Indeed, some of the agencies now contributing expertise and money, such as the National Oceanic and Atmospheric Administration, are high on the list of candidates for political extinction (Science, 19 May, p. 964). "We have the technical knowledge to do the restoration," says Ogden. "But I worry about sustaining the political will."

A RIVER OF GRASS

In the late 1800s, when fewer than 1000 people lived in what are now Dade, Broward, and Palm Beach counties, water spilled over the banks of Lake Okeechobee in the wet season and flowed lazily southward to Florida Bay. This was the "River of Grass," a swath of saw grass and algae-covered water 50 miles wide and only a foot or two deep. People found the vast swamp inhospitable—too wet and too many bugs—but its mosaic of wetland habitats supported a stunning diversity of animals and plants, including huge colonies of wading birds.

Figure 7.1
Everglade snail kite, one of many endangered species that inhabit the Everglades.

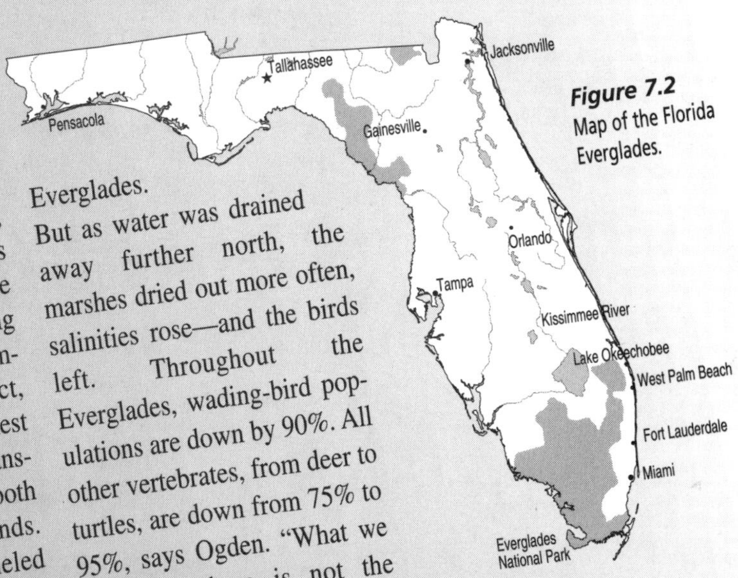

Figure 7.2
Map of the Florida Everglades.

Then the human migration to Florida began. In order to make the River of Grass and adjacent marshlands suitable for cities and agriculture, about half of the Everglades was drained in successive waves of development starting early this century. The mammoth flood-control project, built by the Corps at the behest of the state of Florida, transformed the hydrology of both public and private lands. Today, water is channeled swiftly through 1600 kilometers of canals and 1600 kilometers of levees, stored in parks called "water conservation areas," and partitioned by countless water-control structures. The River of Grass is interrupted by the world's largest zoned farming area, the Everglades Agricul-tural Area (EAA), south of Lake Okeechobee. To prevent flooding, "extra water" is diverted east and west to the Gulf of Mexico and the Atlantic Ocean. The whole system is completely artificial, says Lewis Hornung, a Corps engineer responsible for undoing much of the work of his predecessors on the Kissimmee.

The old Corps engineers recognized that their work would alter the natural world, says Hornung. But no one predicted the devastating effects. For example, hundreds of thousands of birds once nested around the headwaters of the Shark River in the southern Everglades. But as water was drained away further north, the marshes dried out more often, salinities rose—and the birds left. Throughout the Everglades, wading-bird populations are down by 90%. All other vertebrates, from deer to turtles, are down from 75% to 95%, says Ogden. "What we have out there is not the Everglades," he says. "It's a big wet area with spectacular sunsets, but functionally it's not working at all. The animal life in many places is no better than you'd see in roadside ditches in Florida in the summer."

THE BEST LAID PLANS

The good news is that hydrological damage may be reversible, explains ecologist Lance Gunderson of the University of Florida. "It's all there except the water… If we redo the hydrology, it will explode," says Richard Ring, superintendent of Everglades National Park.

But will the flora and fauna come back? Anecdotal reports from marshes in the northern part of the park suggest that the wetlands do indeed revive when freshwater returns, says Steve Davis, senior ecologist at the SFWMD. "And it's sort of common sense," adds Robert Johnson, chief hydrologist at the national park. "Wetlands need to be wet." Still, to date scientists can't cite the results of any large-scale reflooding study to prove this point. Says Ogden: "Hydrological restoration doesn't equal ecological restoration. This is a big uncertainty, and we need to design flexible plans to deal with it."

Plans are already shifting. In late 1994, the Corps released a preliminary study that outlined six alternatives for revamping the hydrology, although they didn't endorse any specific option. Planners now say none of the six is likely to be the solution, admits Stuart Appelbaum, who directs the Corps' Everglades planning process. There's simply no

Continued on next page

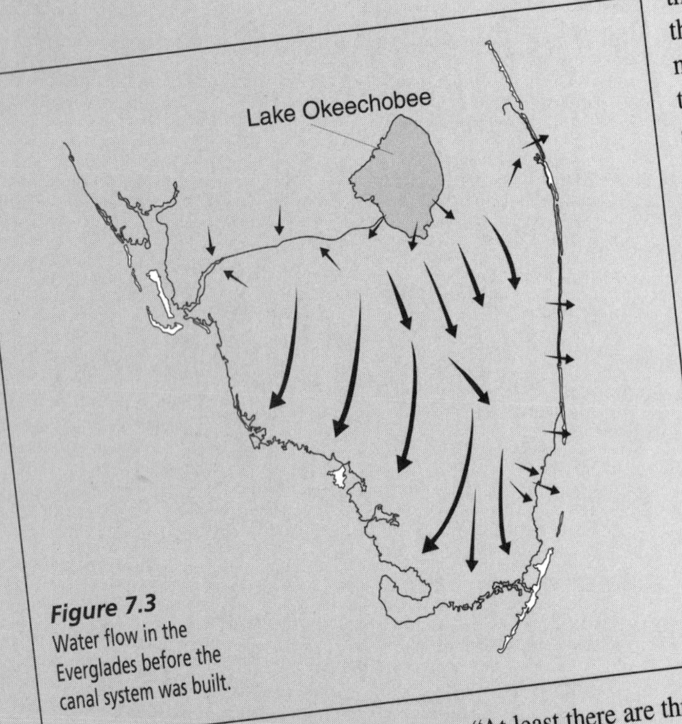

Figure 7.3
Water flow in the Everglades before the canal system was built.

consensus yet on exactly how to increase water storage and flow while guarding against floods. Nor have restorers made tough decisions about which lands to acquire from private owners for water management. The Corps has gone back to its planning; Appelbaum says a coordinated restoration blueprint is due in 6 years.

Frampton and others want a plan sooner. But in the meantime, restorers point to three smaller, independent hydrological efforts that are already entering construction. One is the Kissimmee. A second project will funnel more water to Shark Slough in the northeastern part of Everglades National Park, and a third will create a buffer strip between wetlands and drained crop fields along the park's eastern border. "At least there are three projects you can point to that are more than just words or paper, where things are actually happening," says Colonel Terrence "Rock" Salt, executive director of the federal task force.

These planners are using an approach they call "adaptive management," which basically means learning by doing. For example, as part of reflooding Shark Slough, Hornung's crew needs to move water from one water-conservation area to another. To do so, he could either build a canal—which models say is more efficient—or simply tear down the levee between the areas. He's experimenting by degrading part of the levee and watching what happens.

To researchers, such experiments are nothing out of the ordinary, but admitting that the outcome is unknown is a new idea for engineers accustomed to having a plan and sticking to it, says Salt. "In our legal system (and there have been many lawsuits over the Everglades already), uncertainty is an admission," says Davis of the SFWMD. "And now here we are starting off up front admitting and defining it."

QUALITY CONTROL

One thing ecologists do know is that water quality, as well as quantity, will be a crucial part of any restoration. Reflood the swamps with polluted water, and the historic system is unlikely to return. Says biologist Douglas Morrison of the National Audubon Society in Miami: "You can say, 'Wet it and it will grow'—but then the next question is what will grow?" In the Everglades, the answer is often cattails. These tall plants were once only a small part of Everglades vegetation, cropping up around high-nutrient areas like alligator holes. But today in some places, cattails nearly 4 meters tall completely blanket the wetlands, says ecologist Ronald Jones of Florida International University. "It's a massive conversion at the landscape level," agrees biologist Wiley Kitchens of the National Biological Service in Gainesville, Florida.

The culprit: phosphorus. The historic Everglades had extremely low concentrations of this nutrient, says Jones. Today, extra phosphorus enters the system from the EAA, where water used to irrigate fertilized sugarcane fields picks up a load of phosphorus, then is swiftly channeled to the water-conservation areas. There it spurs nutrient-loving vegetation like cattails and blue-green algae. Jones argues that to be true to the historic system, the Everglades needs very low levels of phosphorus—perhaps as low as 10 parts per billion. "The sugar growers say we want it cleaner than Perrier—and that's true, for phosphorus. That's just the character of the Everglades," he says.

Not surprisingly, the sugar growers are unconvinced. "Ten parts per billion—what's the basis for that? Parts of the Chesapeake Bay watershed are at around 400 ppb," says Peter Rosendahl, vice president of environmental communications at Flo Sun, one of the major sugar companies. He points out that no one really knows how much phosphorus the Everglades can handle; studies are under way now. "There's no real reason to believe that extra nutrients are the cause of the decline in the Everglades," he says.

A partial solution to the problem, one mandated by an act passed by the state legislature last year, calls for a ring of artificial marshes around the EAA to filter phosphorus from the water. A test marsh full of cattails is already up and running.

There are other thorny water quality issues, however. Chief among them is mercury, which is mysteriously contaminating fish and wildlife in the heart of the remote Everglades, to the point that fishers are advised not to eat their catch. So far, no one knows where the mercury is coming from or just how much damage it's causing, says Dan Scheidt, south Florida coordinator at the Environmental Protection Agency. But whether the issue is phosphorus or mercury, it's increasingly clear that specific goals for water quality will have to be addressed in the coordinated restoration plan. "We have some movement on the hydrology," says Salt. "But we haven't yet looked at water-quality issues holistically—and we need to."

SUPPORTING THE SWAMP

The depth of the political backing for the plan also concerns planners. In the current political climate, it's hard to count on ongoing federal commitments. Webb of the Wilderness Society worries that popular support is "like the River of Grass itself—miles wide and only a few inches deep." There's also the small matter of aligning dozens of government agencies and interest groups, from sugar-cane growers to Indian tribes. For example, sugar-cane researcher Barry Glasz of the Department of Agriculture says he doesn't even like the word "restore," because to him it suggests turning the clock back to a time before agriculture. Indeed, many environmental groups would like nothing better than to reduce the sugar industry's presence in South Florida. "The EAA has about half a million acres of sugar. We'd like to see maybe one third of that taken out of production and become wetland or water-retention areas," says Ron Tipton of the World Wildlife Fund.

On the other hand, surveys have shown strong public support for saving the Everglades, says Davis of the SFWMD. And urban planners and utility officials—who want to guard the water supply—agree with environmentalists that some hydrological restoration is needed. In the historic system, wetlands cached rainfall for months and so recharged the ground water of the Biscayne Aquifer, which supplies the thirsty cities of Florida's southeast coast, explains Tom Teets, water supply planner for the SFWMD. Now much of the rainfall is shuttled out to sea long before it seeps into the ground. Water supplies are adequate for the 4.1 million people who lived in Florida's urban southeast coast in 1990, but Teets and others worry about the 6 million

Continued on next page

expected to live there by 2010. "We get 60 inches of rainfall, but we can't retain it because the water has been managed poorly," says Jorge Rodriguez, deputy director of the Miami-Dade Water and Sewer Department. "So we feel everyone can benefit from restoration."

Adjacent to the test fill in the central Kissimmee, water is once again flowing through the ancient oxbow turns. The area affected is too small to see a large influx of wildlife, says Louis Toth of the SFWMD, the Kissimmee's resident biology expert. But vegetation is slowly colonizing the filled-in canal, and game fish are spawning in the newly restored flood plain. Whether uncertain science and precarious political support can engineer a similar recovery for the whole Everglades, however, is still too far downstream to see clearly.

Glossary of Terms

The following terms can be found on the listed page in the Student Manual unless otherwise noted. ◆ indicates pages that you may receive from your teacher.

Term	Page
abiotic	4
adaptations	50
autotroph	5
biodiversity	67
biogeochemical cycles	23
biotic	4
biotic potential	36
calorie	24
carnivore	23
carrying capacity	38
community	15
consumers	17
decomposer	5
denitrification	30
ecology	1
ecosystem	4
endangered species	◆111
evolution	47
exponential growth	36
extinction	52
food chain	17
food web	18
fossil record	51
gradualism	51
habitat	1
herbivore	23
heterotroph	5
J-shaped curve	36
limiting factor	39
natural selection theory	47
niche	15
nitrification	29
nitrogen-fixing	29
photosynthesis	23